ZEITSCHRIFT FÜR ALLGEMEINE MIKRO-BIOLOGIE

AN INTERNATIONAL JOURNAL ON MORPHOLOGY, PHYSIOLOGY, GENETICS, AND ECOLOGY OF MICROORGANISMS

BAND 23 · 1983 · HEFT 4

AKADEMIE-VERLAG · BERLIN

ISSN 0044-2208 Z. Allg. Mikrobiol., Berlin 23 (1983) 4, 209–280 EVP 20,– M

ZEITSCHRIFT FÜR ALLGEMEINE MIKROBIOLOGIE

VOLUME 23 1983 NUMBER 4

CONTENTS

ZEITSCHRIFT FÜR ALLGEMEINE MIKRO-BIOLOGIE

AN INTERNATIONAL JOURNAL ON MORPHOLOGY, PHYSIOLOGY, GENETICS, AND ECOLOGY OF MICROORGANISMS

BAND 23 · 1983 · HEFT 4

AKADEMIE-VERLAG · BERLIN

Die Zeitschrift für Allgemeine Mikrobiologie soll dazu beitragen, Forschung und internationale Zusammenarbeit auf dem Gebiet der Mikrobiologie zu fördern. Es werden Manuskripte aus allen Gebieten der allgemeinen Mikrobiologie veröffentlicht. Arbeiten über Themen aus der medizinischen, landwirtschaftlichen, technischen Mikrobiologie und aus der Taxonomie der Mikroorganismen werden ebenfalls aufgenommen, wenn sie Fragen von allgemeinem Interesse behandeln.

Zur Veröffentlichung werden angenommen:

Originalmanuskripte, die in anderen Zeitschriften noch nicht veröffentlicht worden sind und in gleicher Form auch nicht in anderen Zeitschriften erscheinen werden. Der Umfang soll höchstens $1^1/_2$ Druckbogen (24 Druckseiten) betragen. Bei umfangreicheren Manuskripten müssen besondere Vereinbarungen mit der Schriftleitung und dem Verlag getroffen werden. Kurze Originalmitteilungen über wesentliche, neue Forschungsergebnisse. Umfang im allgemeinen höchstens 3 Druckseiten. Kurze Originalmitteilungen werden beschleunigt veröffentlicht.
Kritische Sammelberichte und Buchbesprechungen nach Vereinbarung mit der Schriftleitung.

Bezugsmöglichkeiten der Zeitschrift für Allgemeine Mikrobiologie:

Bestellungen sind zu richten
— **in der DDR** an den Postzeitungsvertrieb, an eine Buchhandlung oder an den Akademie-Verlag, DDR-1086 Berlin, Leipziger Str. 3—4
— **im sozialistischen Ausland** an eine Buchhandlung für fremdsprachige Literatur oder an den zuständigen Postzeitungsvertrieb
— **in der BRD und Berlin (West)** an eine Buchhandlung oder an die Auslieferungsstelle KUNST UND WISSEN, Erich Bieber OHG, D-7000 Stuttgart 1, Wilhelmstraße 4—6
— **in den übrigen westeuropäischen Ländern** an eine Buchhandlung oder an die Auslieferungsstelle KUNST UND WISSEN, Erich Bieber GmbH, CH-8008 Zürich Schweiz, Dufourstraße 51
— **im übrigen Ausland** an den Internationalen Buch- und Zeitschriftenhandel; den Buchexport, Volkseigener Außenhandelsbetrieb der Deutschen Demokratischen Republik, DDR-7010 Leipzig, Postfach 160, oder an den Akademie-Verlag, DDR-1086 Berlin, Leipziger Straße 3—4.

Zeitschrift für Allgemeine Mikrobiologie
Herausgeber: Im Auftrag des Verlages von einem internationalen Wissenschaftlerkollektiv herausgegeben.
Verlag: Akademie-Verlag, DDR-1086 Berlin, Leipziger Straße 3—4; Fernruf: 2236229 oder 2236121;
Telex-Nr.: 114420; Bank: Staatsbank der DDR, Berlin, Kto.-Nr.: 6836-26-20712.
Chefredaktion: Prof. Dr. UDO TAUBENECK
Prof. Dr. WILHELM SCHWARTZ.
Anschrift der Redaktion: Zentralinstitut für Mikrobiologie und experimentelle Therapie der Akademie der Wissenschaften, DDR-6900 Jena, Beutenbergstr. 11; Fernruf: Jena 885614; Telex-Nr.: 058621
Veröffentlicht unter der Lizenznummer 1306 des Presseamtes beim Vorsitzenden des Ministerrates der Deutschen Demokratischen Republik.
Gesamtherstellung: VEB Druckerei „Thomas Müntzer", DDR-5820 Bad Langensalza.
Erscheinungsweise: Die Zeitschrift für Allgemeine Mikrobiologie erscheint jährlich in einem Band mit 10 Heften. Bezugspreis je Band 250,— M zuzüglich Versandspesen (Preis für die DDR 200,— M), Preis je Heft 25,— M (Preis für die DDR 20,— M).

Erscheinungstermin: Juni 1983

Bestellnummer dieses Heftes: 1070/23/4

AN (EDV) 75218

Zeitschrift für Allgemeine Mikrobiologie	23	1983	4	211–218

(Sciences et Technologies des Industries Alimentaires, Université des Sciences et Techniques du Languedoc, 34060 Montpellier Cedex France, and Laboratoire de Génétique et Microbiologie, INRA-ENSAM, 34060 Montpellier Cedex[1], France)

Selection of yeasts for single cell protein production on media based on Jerusalem artichoke extracts

VÉRONIQUE APAIRE, J. P. GUIRAUD and P. GALZY[1]

(Eingegangen am 20.7.1982)

Several yeast strains can grow with good yield (0.16 to 0.19 mg protein/mg carbohydrate) on nitrogen supplemented Jerusalem artichoke extract. The most promising strain is *Lipomyces starkeyi*. Including by-products (pulps, proteins of extract), protein production can reach 2 metric tons/ha.

The production of single cell protein from industrial wastes, agricultural products or by-products is not only interesting for developing countries but also for industrialized countries which have to supplement animal feed with some sources of protein.

The product is competitive only when the substrate used is cheap and readily available. That is why we investigated the possibility of protein production from Jerusalem artichoke (*Helianthus tuberosus*) extracts, containing polyfructosans of the inulin type. Jerusalem artichoke is a rustic, robust plant which grows well on average quality soil, with good yields (30 to 60 metric tons/hectare). At the present time, Jerusalem artichoke is only used as animal fodder. However, several industrial applications based on this plant have been considered: ethanol production (UNDERKOFLER *et al.* 1937, BOINOT 1942, GUIRAUD *et al.* 1981), fructose syrup (KIERSTAN 1978, FLEMMING and GROOTWASSINK 1979, GUIRAUD and GALZY 1981 a, b).

Several yeast strains adapted to the production of protein on synthetic media with added inulin were selected previously (GUIRAUD *et al.* 1979). Our aim is to show that these yeasts are able to grow on Jerusalem artichoke extract with good protein yield.

Materials and methods

Biological material: We used Jerusalem artichoke of the "Violet de Rennes" variety. The yeast strains were either from the Centraal Bureau voor Schimmelcultures, Delft, Holland (CBS) or from the collection of our own laboratory (LG): *Kluyveromyces fragilis* CBS 1555, *Kluyveromyces marxianus* LG, *Debaryomyces cantarellii* CBS 4349, *Candida salmanticensis* CBS 5121, *Pichia polymorpha* CBS 186, *Lipomyces starkeyi* LG, *Schwanniomyces castellii* CBS 2863.

Cultivation technique: Unless stated otherwise, the cultures were incubated at 28 °C in ERLENMEYER flasks, filled to 1/10 of their volumes and shaked (1.3 Hz; 75 mm amplitude).

Jerusalem artichoke extract preparation: The tubers were washed, diced into 1 cm cubes and crushed with a blade grinder. Water was added to obtain an extract with 650 g of wet matter per liter and the mixture was crushed again with an Ultraturrax, pressed and filtered through cloth. The tyndallization was performed by 3 successive heatings at 24 h interval to 80 °C during 15 min. The extract was sometimes supplemented with the following solution, containing per liter: Ammonium sulfate 60 g, potassium phosphate 50 g, magnesium sulfate 5 g, sodium chloride 1 g, calcium chloride 1 g, boric acid 50 mg, manganese sulfate 40 mg, zinc sulfate 40 mg, sodium molybdate 20 mg, copper sulfate 4 mg.

Analytical techniques: Total sugars were determined by the anthrone method (Trevelyan and Harrison 1982). Nitrogen content was assayed by Kjeldahl method (Bradstreet 1965) and cellular protein by biuret method according to Stickland (1951). Amino acids content was determined with an autoanalyzer after hydrolysis according to the method of Gundlach *et al.* (1959); tryptophan was determined according to Simpson *et al.* (1976). Chemical indices were determined according to Block and Mitchell (1946). The cell free lipid extract was prepared by the azeotropic method (Kahane and Regord 1964) and purified according to Folch *et al.* (1957).

Cellular growth determination was performed by nephelometric measurements at 400 nm. Cellular dry matter was determined by dessication at 104 °C during 24 h.

Results

Characteristics of the extract utilized

We have previously published (Guiraud *et al.* 1981) the average composition of Jerusalem artichoke tubers, extract and resulting pulp after crushing and pressing.

The raw extract used in this experiment contained 150 g of dry matter per liter, 86% of which was carbohydrate and 1.8% was nitrogen, mostly from protein (1.15% protein nitrogen). Since this carbohydrate concentration was much too high for the raw extract to be used directly as culture medium, we diluted the extract to obtain a juice containing 5 g of sugar per liter.

The non-protein nitrogen content of this juice was too low (40 mg/l) for a good protein yield. To obtain 16 to 20 g cellular protein per 100 g of sugars, 120 to 160 mg of nitrogen per liter is needed. It was thus necessary to supplement the juice with at least 160 mg inorganic nitrogen per liter. This supplementation was made with a mineral salt solution so that 1 g of nitrogen per liter medium was obtained. This concentration was clearly higher than minimum requirement. Cultures grown in juice without supplementation served as controls.

Growth studies on tyndalized juice

Our first experiments were carried out with tyndalized juice in order to limit accidents due to eventual contaminations. This medium was used to study the growth of 7 strains previously selected on synthetic media (Table 1).

Growth rates were higher on supplemented juice than on raw juice. The best results were obtained for *Kluyveromyces fragilis* and *Pichia polymorpha* ($\mu = 0.30$). Growth rates were generally higher on Jerusalem artichoke juice than on synthetic medium. This could be due to the lesser degree of carbohydrate polymerization in Jerusalem artichoke juice compared to pure inulin, so that the hydrolysis phase preceding carbohydrate utilization was shorter.

With raw juice, the growth yields, and particularly protein yields, were low. This is expected since the medium was deficient in nitrogen. However, it is interesting to note that some protein yields reached 0.1 mg/mg carbohydrate when non-protein nitrogen content of the juice was expected to allow only yields of up to 0.05 mg protein/mg carbohydrate. This may be explained either by the utilization of part of juice protein by the yeast, or by the recovery of this protein along with the yeast cells during centrifugation. With nitrogen supplementation, growth rates were considerably higher. The increase was particularly remarkable in protein yields which doubled in certain cases. The best dry matter yield was obtained for *Schwanniomyces castellii*, followed by *Lipomyces starkeyi* and *Candida salmanticensis*. The best yield in protein was obtained for *Lipomyces starkeyi*, followed by *Schwanniomyces castellii*. Compared to those obtained from culture on synthetic medium with inulin (Guiraud *et al.* 1979), these results were generally lower except for *Schwanniomyces castellii* and *Lipomyces starkeyi* where they were better.

Table 1

Growth yields of the indicated yeast species after cultivation in tyndalized raw and supplemented Jerusalem artichoke juices (5 g carbohydrate/liter), at normal pH (6.6). R = raw juice, S = supplemented juice, % of total native juice protein

Strain	Media	Growth rate (μ)	Carbohydrate used (%)	Dry matter yield mg DM/mg carbohydrate	Protein yield mg protein/mg carbohydrate	Protein content residual juice %
Candida	R	0.17	96	0.39	0.10	28
salmanticensis	S	0.23	99	0.44	0.16	32
Debaryomyces	R	0.23	95	0,35	0.10	25
cantarellii	S	0.29	97	0.38	0.15	29
Kluyveromyces	R	0.24	95	0.29	0.08	36
fragilis	S	0.30	98	0.36	0.14	41
Kluyveromyces	R	0.26	97	0.28	0.08	29
marxianus	S	0.27	98	0.40	0.14	45
Lipomyces	R	0.26	92	0.35	0.09	34
starkeyi	S	0.27	98	0.48	0.19	50
Pichia	R	0.23	94	0.35	0.07	50
polymorpha	S	0.30	98	0.33	0.11	49
Schwanniomyces	R	0.25	96	0.38	0.10	20
castellii	S	0.27	96	0.48	0.18	23

In each experiment we determined the protein content of the juice after cell recovery. These substances were native proteins already present in the juice but not utilized or recovered with the yeast cell. The levels obtained were low and represented at most 50% of native proteins. We expected a marked difference between supplemented and raw juices since the presence of inorganic nitrogen would limit the utilization of juice proteins. The observed differences for most strains were small. There was probably protein recovery by precipitation during yeast cells recovery.

Table 2

Growth yields on raw and supplemented Jerusalem artichoke juices (5 g carbohydrate/liter, pH ajusted to 3.5 and non-sterilized) R: raw juice, S: supplemented juice, % of total native juice protein

Strain	Media	Growth rate (μ)	Carbohydrate used	Dry matter yield mg DM/mg carbohydrate	Protein yield mg protein/mg carbohydrate	Protein content residual juice %	Protein (and N) content of cells mg/mg DM
Candida	R	0.15	93	0.34	0.09	32	25.5 (4.3)
salmanticensis	S	0.20	99	0.37	0.16	78	42.5 (7.9)
Debaryomyces	R	0.21	96	0.35	0.08	46	22.5 (3.8)
cantarllii	S	0.27	97	0.39	0.14	38	35.5 (6,1)
Kluyveromyces	R	0.24	97	0.29	0.09	35	32 (5.5)
fragilis	S	0.30	98	0.35	0.13	60	38.5 (6,7)
Kluyveromyces	R	0.27	97	0.28	0.10	45	34.5 (5.6)
marxianus	S	0.27	97	0.38	0.14	62	36.5 (6.1)
Lipomyces	R	0.25	88	0.29	0.10	40	34.5 (5.7)
starkeyi	S	0.30	97	0.45	0.17	51	39 (6.5)
Pichia	R	0.23	95	0.31	0.07	28	21.5 (3.8)
polymorpha	S	0.24	96	0.32	0.10	26	31.5 (5.7)
Schwanniomyces	R	0.20	91	0.31	0.08	38	26.5 (4.4)
castellii	S	0.23	96	0.45	0.16	39	35.5 (6)

Growth studies on non-sterile juice

A sterilization process of the juice represents an important production cost factor. Therefore we tried to use a non-sterile juice, acidified to pH 3.5; this pH prevents the development of numerous bacterial contaminants without affecting yeasts.

The growth yields were slightly lower than those observed for tyndallized juice (Table 2). The highest dry-matter yields were obtained for *Schwanniomyces castellii* and *Lipomyces starkeyi*, and the highest protein yields were obtained for *Lipomyces starkeyi*. It should be noted that, except for *Pichia polymorpha*, the protein contents of residual juice were higher at pH 3.5 than at normal pH. The total protein recovery was more efficient at pH 3.5.

Cell composition studies

The protein and KJELDAHL nitrogen contents of the different strains grown on acidified juice (pH 3.5) are summarized in Table 2. The highest protein content was obtained for *Candida salmanticensis* (42.5%), followed by *Lipomyces starkeyi* (39%).

The aminograms for proteins of the different strains were determined after culture on synthetic medium YNB with inuline (Table 3). This medium was chosen so as to limit interference due to recovery of juice proteins. An aminogram was also obtained for *Lipomyces starkeyi*, the most promising strain, grown on Jerusalem artichoke juice. Chemical indices were calculated for the essential amino acids.

All the strains have a good composition of essential amino acids (C.I. = 100), except for the sulphur-containing amino acids methionine and cysteine. *Schwanniomyces castellii* and fortunaly, *Lipomyces starkeyi* were characterized by satisfactory contents of all essential amino acids including methionine and cysteine. This is fairly rare among yeasts which generally are poor in sulphur-containing amino acids. *Lipomyces starkeyi* grown on Jerusalem artichoke juice showed a different aminogram, especially with regard to sulphur-containing amino acids which were clearly deficient. This confirmed that juice proteins were recovered along with yeast cells during the centrifugation step.

Since *Lipomyces starkeyi* strain is reputed to have a high lipid content (STARKEY 1946), it seemed interesting to verify this property after cultivation on supplemented Jerusalem artichoke juice at 5 g carbohydrate/liter. The lipid content of dry matter was found to amount to 12.5%. This value was considerably higher than the lipid contents (5—10%) of various yeast strains reported in the literature (KORENAGA 1977, MOULIN *et al.* 1975).

Attempts to optimize protein production

The protein yields obtained for the 7 strains used in these experiments were overestimated since some juice protein was recovered along with the yeast cells. Under these conditions, it might be interesting to recover juice protein before yeast inoculation. When a Jerusalem artichoke juice was adjusted to pH 3.5 and treated to 90 °C, about 40% of the proteins precipitated and could be recovered by centrifugation. The protein content of a juice submitted to such a treatment was reduced from 10.4 to 6.5 g protein/l. We have grown *Lipomyces starkeyi* cultures on such a juice. The protein yield thus obtained was 0.145, which was lower than with raw juice (0.175); at the same time about 10% decrease of protein that obtained content in residual juice was observed. The total amounts of protein recovered by the two methods were about the same.

The deproteinization technique before culture inoculation required a lot of energy and seemed not to have any advantage. Ultrafiltration technique would also be dif-

Table 3
Aminogram of proteins of different yeast species after incubation in YNB-inuline medium or Jerusalem artichoke juice. The amino acid contents are expressed as % of total protein. Chemical indices of essential amino acids are indicated in parentheses

Amino acid	YNB inulin medium							Jerusalem artichoke juice
	Candida salmanticensis	*Debaryomyces cantarellii*	*Kluyveromyces fragilis*	*Kluyveromyces marxianus*	*Lipomyces starkeyi*	*Pichia polymorpha*	*Schwanniomyces castellii*	*Lipomyces starkeyi*
Asp.	11.0	11.7	10.5	10.6	10	10.6	11.6	10.3
Thre.	6.4 (160)	5.7 (141)	6 (150)	5.6 (139)	5.3 (133)	6.7 (168)	5.4 (136)	5.9 (148)
Ser.	6.6	6.3	6	5.2	5.2	8.1	5.8	5.9
Glut.	13.8	17.5	13.3	16.3	17.5	10.8	13.4	12.7
Pro.	3.7	3.3	4.0	3.7	3.6	3.8	4.1	4
Gly.	5.2	5.1	4.9	4.7	4.8	6.2	4.8	5.4
Ala.	7.4	6.0	9.9	9.1	5.4	5.7	5.6	8.1
Cyst.	1.2	1.5	1.3	1.3	1.6	1.3	1.5	0.7
Met.	1.4 (75)	1.6 (88)	1.6 (84)	1.6 (83)	1.8 (97)	1.7 (85)	1.8 (97)	1.1 (52)
Val.	6.1 (122)	5.4 (108)	6.1 (122)	5.3 (106)	5.1 (102)	6.0 (121)	5.5 (110)	5.8 (116)
Ileu.	4.7 (117)	4.9 (121)	4.6 (115)	4.1 (104)	4.1 (102)	5.1 (127)	5.1 (128)	5 (125)
Leu.	7.7 (110)	7.5 (108)	7.5 (108)	7.6 (109)	6.9 (99)	7.7 (110)	7.8 (112)	8 (114)
Tyr.	3.5	3.5	3.7	3.9	3.2	4.3	3.6	3.9
Phe.	4.6 (135)	4.3 (130)	3.7 (122)	4.7 (144)	3.6 (114)	4.6 (148)	4.7 (138)	4.9 (147)
His.	2.2	2.1	2.3	2.2	2.1	2.4	2.6	2.6
Lys.	7.4 (135)	7.4 (134)	8.0 (145)	7.9 (143)	7.8 (141)	8.6 (156)	8.4 (152)	7.6 (138)
Arg.	5.4	4.5	4.6	4.9	10	4.6	6.3	5.2
Try.	1.5 (143)	1.7 (166)	2.0 (197)	1.4 (144)	1.3 (140)	1.9 (185)	1.8 (175)	2.2 (220)

ficult to apply due to the viscosity imparted to the juice by its carbohydrates. It is easier to recover by ultrafiltration the residual proteins of the juice after cultivation and harvest the yeast cells by centrifugation. When *Lipomyces starkeyi* was grown on supplement juice acidfied to pH 3.5, we recovered by this method 395 mg of residue from 1 liter of medium. This residue contained 49% proteins and 8% nitrogen. The aminogram of these proteins is given in Table 4. Leucine, isoleucine and lysine contents were low. After this operation, the remaining juice contained 400 mg dry matter per liter, with 25% carbohydrates and 60% ash. Its chemical oxygen demand was 350 mg O_2 per liter.

Table 4
Aminogram of proteins from pulp and ultrafiltrate residue after cultivation of *Lipomyces starkeyi*. The amino acid contents are expressed as % of total protein. Chemical indices of essential amino acids are indicated in parentheses.

Amino acid	Pulp	Ultrafiltrate residue
Asp.	10	13
Thre.	5.6 (140)	9.2 (230)
Ser.	5.6	8.7
Glut.	12	15.5
Pro.	5.2	4.8
Gly.	5.2	7.2
Ala.	5.6	5.4
Cyst.	1.3 (63)	2.6 (110)
Met.	0.9	1.3
Val.	6.2 (124)	5.3 (106)
Ileu.	4.7 (117)	3.4 (85)
Leu.	8.5 (121)	4.3 (61.5)
Tyr.	3 (133)	3.6 (108)
Phe.	5	2.9
His.	2.4	2.2
Lys.	9 (163)	5.3 (96)
Arg.	12.5	2.5
Try.	2.3 (230)	1.8 (180)

The pulp also constituted an interesting by-product since it contains proteins (7%) and can be used as animal feed. The aminogram of these proteins was also determined and is presented in Table 4. Similarly to yeasts there was a deficiency in sulphur-containing amino acids.

It also seemed interesting to perform culture experiments with *Lipomyces starkeyi* with varying supplementation levels in order to determine the minimal requirements. The minimal mineral salt concentration allowing the optimal protein yield of 0.174 mg/mg carbohydrates corresponded to a supplementation of 200 mg nitrogen/l. In fact, nitrogen seemed not to be the only limiting element since the suppression of other elements caused a 10% decrease in yield. We have also successfully performed continuous cultivation with supplemented juice acidified to pH 3.5 without contamination problems.

Discussion

Jerusalem artichoke juice can be used as substrate for different yeast strains, however, it must be supplemented with an ammonium salt in order to obtain the maximum yield. The yeast strain having the best growth rate, growth yield and protein content is *Lipomyces starkeyi*. *Schwanniomyces castellii* also has a good protein yield and a satisfactory growth rate, however, its protein content is lower. The essential amino acid contents of the strains tested are satisfactory, however, there is generally a slight deficiency in sulphur-containing amino acids which is a well-known fact among yeasts. Both *Lipomyces starkeyi* and *Schwanniomyces castellii* contain satisfactory levels of cysteine and methionine. These levels were less satisfactory in cells grown on Jerusalem artichoke juice as medium.

Since yeast proteins are never used above 20% in the rations because of toxicity problems due to high nucleic acid content, the partial deficiency can be easily complemented in the diet by other sources of proteins, as for instance cereals. The low price of DL-methionine would allow also an economical supplementation of the different products obtained.

Considering its performance, *Lipomyces starkeyi* is quite appropriate for protein production. *Lipomyces starkeyi* production also presents another interesting feature: its lipids content is good. This may enable one to extend the use of Jerusalem artichoke juice to the biological production of fats and oils.

Total protein production from juice can be estimated to be 30 kg/metric ton of fresh Jerusalem artichoke. Taking into account the pulp (1 to 3 metric tons/ha), the total becomes 36 kg protein/metric ton of Jerusalem artichoke or 1 to 2 metric tons protein/ha. The production of food yeast from Jerusalem artichoke extracts can also become interesting in the prospect of using by-products of other industrial processes such as the production of ethanol or sugar. Waste waters and solutions, too, diluted for the principal production may be used with advantage for protein production.

References

BLOCK, R. J. and MITCHELL, H. H., 1946. The correlation of the amino acids composition of proteins with their nutritive value. Nutr. Abst. Rev., **16**, 696–701.

BOINOT, F., 1942. Le topinambour dans la fabrication de l'alcool. Bull. Ass. Chim. Fr., **59**, 792–805.

BRADSTREET, R. B., 1965. The KJELDAHL Method for Organic Nitrogen. Academic Press New York.

FLEMMING, S. E. and GROOTWASSINK, J. W. D., 1979. Preparation of high fructose syrup from the tubers of the Jerusalem artichoke. CRC Crit. Rev. Food. Sci. Nutr., **11**, 1–28.

FOLCH, J., LEES M. and SLOANE-STANLEY, G. M., 1957. Isolation and purification of total lipids from tissues. J. biol. Chem., **226**, 497–504.

GUIRAUD, J. P., DEVOUGE, C. and GALZY, P., 1979. Selection of yeast strains with a view to the production of proteins on inulin. Biotechnol. Letters, **1**, 461–464.

GUIRAUD, J. P. et GALZY, P., 1981a. Production de fructose par hydrolyse chimique de l'inuline. Ind. Alim. Agric., **98**, 45–52.

GUIRAUD, J. P. and GALZY, P., 1981b. Study of enzymatic hydrolysis of plant extracts containing inulin. Enz. Microbiol. Technol., **3**, 305–308.

GUIRAUD, J. P., DAURELLES, J. and GALZY, P., 1981. Alcohol production from Jerusalem artichoke using yeasts with inulinase activity. Biotechnol. Bioeng., **23**, 1461–1465.

GUNDLACH, H. G., MOORE, S. and STEIN, W. J., 1959. Reaction of iodoacetate with methionine. J. biol. Chem., **243**, 1761–1766.

KAHANE, E. et REGORD, M. T. 1964. Sur l'extraction des lipides. Etude des levures. Ann. Nutr. Alim., **18**, 1–4.

KIERSTAN, M. P. J., 1978. Production of fructose syrups from inulin containing plants. Biotechnol. Bioeng., **20**, 447–450.

KORENAGA, H., 1977. Determination of simple defined media for normal growth and lipid production of the yeast *Lipomyces starkeyi*. Nippon Nogei Kagakukaishi, **51**, 449–457.

MOULIN, G., RATOMAHENINA, R., GALZY, P. and BEZARD, J., 1975. Relationships between the presence of linolenic acid and the ability to form respiration deficient mutants in yeast. Folia Microbiol., **20**, 396–401.

SIMPSON, R. J., NEUBERGER, M. R. and LIU, P. Y., 1976. Complete amino acid hydrolysis of proteins from a single hydrolysate. J. biol. Chem., **251**, 1936–1941.

STARKEY, R., 1946. Lipid production by a soil yeast. J. Bacteriol., 51–57.

STICKLAND, L. H., 1951. The determination of small quantities of bacteria by means of the biuret reaction. J. gen. Microbiol., **5**, 698—703.
TREVELYAN, W. E. and HARRISON, J. S., 1956. Studies on yeast metabolism. Biochem. J., **63**, 23 to 28.
UNDERKOFLER, L. A., MCPHERSON, W. K. and FULMER, E. I., 1937. Alcoholic fermentation of Jerusalem artichoke. Ind. Eng. Chem., **29**, 1160—1165.

Mailing address: Dr. P. GALZY
Laboratoire de Génétique et Microbiologie
INRA-ENSAM
Place Viala
34060 Montpellier Cedex, France

Zeitschrift für Allgemeine Mikrobiologie	23	1983	4	219–224

(Sektion Biologie der Ernst-Moritz-Arndt-Universität Greifswald,
WB Molekularbiologie)

Gen-Enzym-Beziehungen des *arom*-Aggregates von *Schizosaccharomyces pombe*

R. Bode

(Eingegangen am 20. 8. 1982)

The gene-enzyme relationships of the *arom* multienzyme complex of *Schizosaccharomyces pombe* that catalyzes steps two through six in the prechorismate polyaromatic amino acid biosynthetic pathway have been studied. The various mutants were subjected to biochemical analysis by direct enzymic assays. These studies have established that *aro*-3A, *aro*-3B, *aro*-3C, *aro*-3D, and *aro*-3E mutants lack, respectively, the enzymic activities 5-dehydroquinate synthase, 5-dehydroquinase, shekimate kinase, 3-enolpyruvylshikimate 5-phosphate synthase, and shikimate: NADP oxidoreductase. In *S. pombe* lack enzymic activities for the inducible quinate catabolic pathway. The functional significance of the *arom* aggregate is discussed.

Die Biosynthese der aromatischen Aminosäuren zeichnet sich durch eine unterschiedliche Organisation der beteiligten Enzyme bei den verschiedenen Organismen aus. Dies trifft sowohl für die vom Chorismat ausgehende Tryptophan- (Crawford 1975) und Phenylalanin/Tyrosin-Synthese (Gibson u. Pittard 1968) zu als auch für die Biosyntheseschritte zur Bildung von Chorisminsäure aus Erythrose-4-phosphat und Phosphoenolpyruvat.

Die Biosyntheseenzyme des Shikimat-Pathway 5-Dehydrochinat-Synthase (DHQ-Synthase), 5-Dehydrochinase (DHQase), Shikimat-Dehydrogenase (Shikimat-DHG), Shikimat-Kinase und 3-Enolpyruvylshikimat-5-phosphat-Synthase (EPSP-Synthase) von Bakterien (Berlyn u. Giles 1969), einiger photosynthetischer Organismen (Berlyn *et al.* 1970) und einzelner Hefen (Bode u. Birnbaum 1981 a) lassen sich voneinander trennen. Dagegen konnte in einigen Pilzen (Ahmed u. Giles 1969, Bode u. Birnbaum 1981 a) und *Euglena gracilis* (Patel u. Giles 1979) nachgewiesen werden, daß diese Enzyme als multifunktionelles Enzymaggregat assoziiert sind.

Die Organisation dieses *arom*-Aggregates konnte bei *Neurospora crassa* (Giles *et al.* 1967) und *Saccharomyces cerevisiae* (De Leeuw 1967) aufgeklärt werden. Strauss (1979) konnte für *Schizosaccharomyces pombe* ebenfalls einen Komplexlocus (*aro* 3) für dessen Synthese nachweisen.

Die Klärung der Gen-Enzym-Beziehungen dieses multifunktionellen Aggregates bei *S. pombe* steht im Vordergrund dieser Arbeit.

Material und Methoden

Stämme und Wachstumsbedingungen: Für die Untersuchungen wurden die *aro* 3-Mutanten 3A-86, 3B-168, 3C-152, 3D-188 und 3E-63 (Strauss 1979) sowie der Wildstamm 972 (h^-) von *Schizosaccharomyces pombe* verwendet. Wir danken Prof. P. Thuriaux, Institut für Allgemeine Mikrobiologie der Universität Bern, Schweiz, für die Überlassung der Stämme. Die Anzucht der Hefen erfolgte als Schüttelkultur bis zum Erreichen der stationären Phase bei 30 °C. Als Nährlösung wurde das Minimalmedium nach Gutz *et al.* (1974) mit 1% Glutaminsäure als Stickstoffquelle verwendet (Strauss 1979). Das Nährmedium für die *aro* 3-Stämme wurde mit 10^{-3} M L-Phenylalanin, L-Tyrosin, L-Tryptophan und 10^{-5} M *p*-Aminobenzoesäure supplementiert.

Enzymextrakt-Gewinnung: Der Aufschluß der Hefezellen erfolgte mit Hilfe einer X-Presse in 0,1 M Tris-HCl (pH 7,5). Das Homogenat wurde bei 20000 g 20 min zentrifugiert und der Überstand zur Bestimmung der Enzymaktivität verwendet. Zur Ermittlung der Inhibitorwirkung einiger Effektoren wurde der Enzymextrakt vorher einer Gelfiltration über Sepharose 6B (2 × 45 cm) unterzogen. Die Säule wurde mit 0,1 M Tris-HCl (pH 8,9) bzw. 0,1 M K-Phosphat-Puffer (pH 7,4) zur Untersuchung der Shikimat-DHG bzw. der Shikimat-Kinase äquilibriert und Fraktionen von 2 ml gesammelt.

Bestimmung der Enzymaktivitäten: Die Aktivitäten der 5-Dehydrochinat-Synthase, 5-Dehydrochinase (EC 4.2.1.10), Shikimat-Dehydrogenase (EC 1.1.1.25), Shikimat-Kinase (EC 2.7.1.71), 3-Enolpyruvylshikimat-5-phosphat-Synthase (EC 2.5.1.19) und Chinat-Dehydrogenase (EC 1.1.1.24) wurden nach Bode u. Birnbaum (1981b) nachgewiesen. Zur Bestimmung der Dehydroshikimat-Dehydratase (DHS-Dehydratase) fand die Methode von Case *et al.* (1972) Verwendung.

Proteinbestimmung: Die Proteinkonzentration der Enzymextrakte wurde nach Lowry *et al.* (1951) bestimmt. Rinderserumalbumin diente dabei als Eichprotein.

Ergebnisse

Identifizierung der *aro* 3-Mutanten

Aus den Resultaten der Feinstrukturanalyse, einschließlich Rekombinations-, Komplementations- und nonsens-Suppressions-Untersuchungen, konnte Strauss (1979) den *aro* 3-Locus in die fünf Subregionen A, B, C, D und E unterteilen, die in der angegebenen Reihenfolge transkribiert bzw. translatiert werden.

Unsere Untersuchungen lassen eine eindeutige Zuordnung der fünf Enzymaktivitäten zu diesen *aro* 3-Subregionen zu (Tab. 1). Die Mutante 3A-86 ist defekt für die DHQ-Synthase, 3B-168 für die DHQase, 3C-152 für die Shikimat-Kinase, 3D-188 für die EPSP-Synthase und 3E-63 für die Shikimat-DHG. Die geringere Aktivität der Shikimat-DHG von 3D-188 wird offenbar durch einen polaren Effekt hervorgerufen. Die anderen Mutantenallele bewirken keine Veränderungen der sie umgebenden *aro* 3-Enzymaktivitäten.

Tabelle 1
Spezifische Enzymaktivität (pkat/mg Protein) des Wildstammes und verschiedener *aro* 3-Mutanten von *S. pombe*

Stamm	DHQ-Synthase	DHQase	Shikimat-DHG	Shikimat-Kinase	EPSP-Synthase
972 (WT)	15	20	100	19	7
3A-86	0	16	110	24	7
3B-168	14	0	110	20	5
3C-152	15	16	100	1	6
3D-188	16	19	55	22	0
3E-63	14	20	0	20	6

Aus den erhaltenen Ergebnissen kann abgeleitet werden, daß die Translation mit der Synthese der Enzymaktivität für die DHQ-Synthase beginnt, via DHQase, Shikimat-Kinase, EPSP-Synthase verläuft und mit der Shikimat-DHG endet.

Regulation des *ARO* 3-Proteins

Die spezifische Aktivität der Enzymaktivitäten des *arom*-Aggregates ließ sich durch Anzucht in Gegenwart der drei aromatischen Aminosäuren nicht wesentlich verändern (Tab. 2). Auch der Einsatz von Amitrol als potentiellen Effektor zur Auslösung der allgemeinen Kontrolle (Wolfner *et al.* 1975) konnte die Enzymaktivitäten ebensowenig beeinflussen wie die unter Aminosäure-limitierten Bedingungen gewachsenen

Mutanten 3B-168 und 3E-63. Die Wachstumsrate wurde durch diese Versuche wesentlich verändert. Wir nehmen deshalb an, daß die Synthese des pentafunktionellen *arom*-Aggregates bei *S. pombe* konstitutiv erfolgt.

Tabelle 2

Spezifische Enzymaktivitäten (pkat/mg Protein) und Wachstumsrate (μ) von *S. pombe* nach unterschiedlichen Wachstumsbedingungen

Stamm	Supplement	Wachstums-rate	DHQ-Synthase	DHQ ase	Shikimat-DHG	Shikimat-Kinase	EPSP-Synthase
972	—	0,28	15	20	100	19	7
972	1 mM Phe/Tyr/Trp	0,24	nd	19	110	20	nd
972	1 mM Amitrol	0,12	nd	22	110	20	nd
3B-168	1 mM Phe/Tyr/Trp	0,20	14	0	110	20	5
3B-168	0,1 mM Phe/Tyr/Trp	0,08	nd	0	120	22	nd
3E-63	1 mM Phe/Tyr/Trp	0,22	14	20	0	20	6
3E-63	0,1 mM Phe/Tyr/Trp	0,09	nd	20	0	18	nd

nd—nicht bestimmt

Eine Inhibition von je 1 mM L-Phenylalanin, L-Tyrosin und L-Tryptophan auf die Enzymaktivitäten des *ARO* 3-Proteins konnte nicht festgestellt werden. Dagegen inhibieren Gallussäure, Vanillin und Brenzkatechin die Shikimat-DHG-Aktivität und Protocatechussäure die Aktivität der Shikimat-Kinase. Andere potentielle Effektoren konnten die Aktivität der beiden in Tabelle 3 dargestellten Enzyme nicht beeinflussen. Die DHQ-Synthase, DHQase und EPSP-Synthase ließen sich durch die aufgelisteten Substanzen nicht in ihrer enzymatischen Aktivität verändern.

Tabelle 3

Relative Aktivität der Shikimat-DHG und der Shikimat-Kinase von *S. pombe* in Gegenwart verschiedener Effektoren [1])

Effektor	Shikimat-DHG	Shikimat-Kinase
Gallussäure	0,46	1,0
Vanillin	1,0	0,91
Protocatechussäure	1,0	1,0
Gentisinsäure	1,0	1,0
Ferualäure	1,0	1,0
Chinasäre	0,43	1,0
Brenzkatechin	1,0	1,0
Phloroglucin	1,0	1,0
Hydrochinon	1,0	1,0
Pyrogallol	1,0	1,0
Resorcin	1,0	1,0

[1]) Enzymansatz für Shikimat-DHG: 1 mM Effektor, 1 mM Shikimisäure, 1 mM NADP, 2 mM Cystein, 0,1 M Tris-HCl (pH 8,9). Enzymansatz für Shikimat-Kinase: 0,5 mM Effektor, 0,5 mM Shikimisäure, 5 mM ATP, 5 mM $MgCl_2$, 10 mM KF, 0,1 M K-Phosphatpuffer (pH 7,4)

Vorkommen eines Chinat-Pathway

Der Vorteil eines *arom*-Enzymaggregates für die Zelle wird oft damit erklärt, daß dieser einen vorhandenen induzierbaren katabolen Chinat-Pathway von dem anabolen Chorisminsäureweg abschirmt, da die Umwandlung von Dehydrochinat zu Dehydroshikimat in beiden Synthesewegen identisch ist (Giles *et al.* 1967, Ahmed u. Giles 1969).

Unsere Untersuchungen an *S. pombe* weisen darauf hin, daß dieser Chinat-Pathway in dieser Hefe nicht existiert. Zum einen können die Zellen nicht mit 0,5% Chinasäure als alleiniger C-Quelle wachsen, zum anderen wird in Gegenwart von Chinasäure zur normalen Energiequelle keine Enzyminduktion beobachtet (Tab. 4). Vom Chinat-Pathway läßt sich praktisch nur die Chinat-DHG in *S. pombe* nachweisen. Eine zweite DHQase-Aktivität konnte mit Hilfe der Mutante 3B-168 nicht festgestellt werden. Ebenso negativ verlief die Bestimmung einer DHS-Dehydratase-Aktivität, deren Aufgabe in der Umwandlung von Dehydroshikimisäure zu Protocatechussäure besteht.

Tabelle 4
Spezifische Aktivität (pkat/mg Protein) der Chinat-DHG, DHQase, Dehydroshikimat-Dehydratase und der Shikimat-DHG von *S. pombe* nach unterschiedlichen Wachstumsbedingungen

Stamm	C-Quelle	Chinat-DHG	DHQase	DHS-Dehydratase	Shikimat-DHG
972	1% Glucose	400	20	0	100
972	0,5% Glucose + 0,5% Chinasäure	320	18	0	90
3B-168	1% Glucose	280	0	0	110
3B-168	0,5% Glucose + 0,5% Chinasäure	320	0	0	100

Diskussion

Der Komplexlocus für die Synthese des *arom*-Proteins kann sowohl in *Neurospora crassa* (Giles *et al.* 1967) und *Saccharomyces cerevisiae* (De Leew 1967) als auch in *S. pombe* (Strauss 1979) in fünf Subregionen unterteilt werden, die die enzymatischen Schritte zwei bis sechs der Biosynthese aromatischer Aminosäuren katalysieren. Durch Untersuchungen des codierten Proteins konnte für alle drei untersuchten Organismen eine pentafunktionelle Proteinkette nachgewiesen werden (Lumsden u. Coggins 1977, De Leeuw 1968, Bode u. Kunze 1983).

Der strukturelle Aufbau des *arom*-Proteins ist bei diesen drei Arten aber nicht identisch. Das *arom*-Protein von *S. pombe* ist in der Reihenfolge DHQ-Synthase—DHQase—Shikimat-Kinase—EPSP-Synthase—Shikimat-DHG codiert; von *N. crassa* in der Reihenfolge DHQ-Synthase—EPSP-Synthase—Shikimat-Kinase—DHQase—Shikimat-DHG (Giles *et al.* 1967). Von De Leeuw (1967) wird die Translationsrichtung für das *Saccharomyces cerevisiae*-Protein mit EPSP-Synthase—Shikimat-Kinase—DHQ-Synthase—Shikimat-DHG angegeben. Der Platz, den die DHQase-Aktivität einnehmen würde, ist noch offen, weil deren Mutanten nicht komplementierten.

Das Gemeinsame der drei Proteine ist, daß ihre Endsequenzen die Shikimat-DHG-Aktivität beinhalten. Im Gegensatz zu *Neurospora crassa* ist die DHQase mit der EPSP-Synthase in der Reihenfolge bei dem *arom*-Aggregat von *S. pombe* ausgetauscht. Die Anordnung der anderen Subregionen ist sonst identisch. Wesentlich größere Differenzen in der Organisation des *arom*-Proteins zu den beiden oben genannten Organismen liegen in *Sacch. cerevisiae* vor.

Die Translationsrichtung der einzelnen Subregionen des *arom*-Locus kann aber von der Reihenfolge der katalytischen Zentren der einzelnen Enzymaktivitäten innerhalb der gebildeten Tertiärstruktur unterschiedlich sein. Größere Differenzen innerhalb der Komplexloci könnten so beim nascenten Protein nivelliert werden. Protein-analytische Methoden müßten darüber Auskunft geben.

Die Vorteile der Ausbildung eines *arom*-Komplexes können für die Zelle darin bestehen, daß ein Channeling der Intermediate der Biosynthese aromatischer Aminosäuren erfolgt. Für viele Pilze dient er offenbar auch dazu, die entstehenden Zwischenverbindungen vor dem Angriff induzierbarer katabolischer Enzyme des Chinat-Pathway abzuschirmen (GILES *et al.* 1967, AHMED u. GILES 1969, BERLYN u. GILES 1972). Für *S. pombe* konnte aber ebensowenig die Existenz dieses Weges nachgewiesen werden wie für *Sacch. cerevisiae* (DE LEEUW 1968), *Euglena gracilis* (PATEL u. GILES 1979) oder *Ustilago violacea* (BERLYN u. GILES 1972).

In *S. pombe* wie auch in den oben genannten Organismen konnte keine Regulierbarkeit der Enzymsynthese des *arom*-Aggregates festgestellt werden. Neben der Sicherung einer koordinierten Bildung der fünf Enzymaktivitäten durch den Komplexlocus, läßt sich noch ein weiterer Unterschied zu den Enzymen des Shikimat-Pathway aufzeichnen, die separabel vorliegen. Die Inhibierbarkeit der Shikimat-DHG und der Shikimat-Kinase durch Phenole und Substanzen, die der Shikimisäure ähnlich sind, ist von *S. pombe* gegenüber Organismen, die kein *arom*-Protein bilden (BALINSKY u. DAVIES 1961, BODE u. BIRNBAUM 1981 b), sehr stark reduziert. Eine nicht mit den Endprodukten koordinierte Regulationsmöglichkeit zur Bildung der aromatischen Aminosäuren wird dadurch vermindert. Eine bisher wenig beachtete Eigenschaft des *arom*-Aggregates, die aber für den Organismus sichtbare Vorteile bringen kann.

Literatur

AHMED, S. I. and GILES, N. H., 1969. Organization of enzymes in the common aromatic synthetic pathway: evidence for aggregation in fungi. J. Bacteriol., **99**, 231—237.

BALINSKY, D. and DAVIES, D. D., 1961. Aromatic biosynthesis in higher plants. 2. Mode of attachment of shikimic acid and dehydroshikimic acid to dehydroshikimic reductase. Biochem. J., **80**, 296—300.

BERLYN, M. B., AHMED, S. I. and GILES, N. H., 1970. Organization of polyaromatic biosynthetic enzymes in a variety of photosynthetic organisms. J. Bacteriol., **104**, 768—774.

BERLYN, M. B. and GILES, N. H., 1969. Organization of enzymes in the polyaromatic synthetic pathway: separability in bacteria. J. Bacteriol., **99**, 222—230.

BERLYN, M. B. and GILES, N. H., 1972. Studies of aromatic biosynthetic and catabolic enzymes in *Ustilago maydis* and in mutants of *U. violacea*. Genet. Res. Camb., **19**, 261—270.

BODE, R. und BIRNBAUM, D., 1981 a. Aggregation und Trennbarkeit der Enzyme des Shikimat-Pathway bei Hefen. Z. Allg. Mikrobiol., **21**, 417—422.

BODE, R. und BIRNBAUM, D., 1981 b. Eigenschaften und Regulation der Enzyme des Shikimat-Pathway von *Hansenula henricii*. Biochem. Physiol. Pflanzen, **176**, 329—341.

BODE, R. und KUNZE, G., 1983. Reinigung und Gewinnung des *arom*-Aggregates von *Schizosaccharomyces pombe*. Z. Allg. Mikrobiol., **23**, 289—296.

CASE, M. E., GILES, N. H. and DOY, C. H., 1972. Genetical and biochemical evidence for further interrelationships between the polyaromatic synthetic and the quinate-shikimate catabolic pathway in *Neurospora crassa*. Genetics, **71**, 337—348.

CRAWFORD, I. P., 1975. Gene rearrangements in the evolution of the tryptophan synthetic pathway. Bacteriol. Rev., **39**, 87—120.

DE LEEUW, A., 1967. Gene enzyme relationships in polyaromatic auxotrophic mutants in *Saccharomyces cerevisiae*. Genetics, **56**, 554—555.

DE LEEUW, A., 1968. Gene-enzyme relationships in aromatic biosynthesis in yeast. Ph. D. thesis, Univ. Yale.

GIBSON, F. and PITTARD, J., 1968. Pathways of biosynthesis of aromatic amino acids and vitamins and their control in microorganisms. Bacteriol. Rev., **32**, 465—492.

GILES, N. H., CASE, M. E., PARTRIDGE, C. W. H. and AHMED, S. I., 1967. A gene cluster in *Neurospora crassa* coding for an aggregate of five aromatic synthetic enzymes. Proc. Natl. Acad. Sci., **58**, 1453—1460.

GUTZ, H., HESLOT, H., LEUPOLD, U. and LOPRIENO, N., 1974. *Schizosaccharomyces pombe*. In: KING, R. C. (ed.), Handbook of Genetics, **1**, 395—445. Plenum Press New York.

LOWRY, O. H., ROSEBROUGH, N. J., FARR, A. L. and RANDALL, R. J., 1951. Protein measurement with the Folin phenol reagent. J. biol. Chemistry, **193**, 265—275.

LUMSDEN, J. and COGGINS, J. R., 1977. The subunit structure of the *arom* multienzyme complex of *Neurospora crassa*. A possible pentafunctional polypeptide chain. Biochem. J., **161**, 599—607.

Patel, V. B. and Giles, N. H., 1979. Purification of the *arom* multienzyme aggregate from *Euglena gracilis*. Biochim. biophysica Acta, **567**, 24–34.
Strauss, A., 1979. The genetic fine structure of the complex locus *aro* 3 involved in early aromatic amino acid biosynthesis in *Schizosaccharomyces pombe*. Molec. Gen. Genet., **172**, 233–241.
Wolfner, M., Yep, D., Messenguy, G. and Fink, G. R., 1975. Integration of amino acid biosynthesis into the cell cycle of *Saccharomyces cerevisiae*. J. Mol. Biol., **96**, 273–290.

Anschrift: Dr. sc. R. Bode
Ernst-Moritz-Arndt-Universität Greifswald
Sektion Biologie, WB Molekularbiologie
DDR-2200 Greifswald, Fr.-Ludwig-Jahn-Str. 15a

Zeitschrift für Allgemeine Mikrobiologie	23	1983	4	225–233

(Central Rice Research Institute, Cuttack-753006, Orissa, India)

Preparation and partial characterization of *Xanthomonas oryzae* phytotoxin

R. S. DUBEY

(Eingegangen am 12. 7. 1982)

A synthetic medium containing ammonium nitrate, urea, D-glucose, glycerol, and sodium glutamate was developed for growth of *Xanthomonas oryzae* and preparation of its phytotoxin. Culture filtrates and partially purified toxin preparations caused blight symptoms when inoculated in leaves of 30 days old rice plant cv. T(N)1. Toxin elicited quicker response than living cells. This activity was due to a non-dialysable, partial acid and heat labile glycopeptide. There was a sharp decrease in activity when culture filtrates and crude toxin preparations were heated at 100 °C for 30 min. Maximum activity was observed in pH range of 6.0–7.0 and a gradual loss at lower pH (5.0–3.0) and higher pH values (8.0–10.0). Unlike living cells which caused continuous lesion enlargement, toxin inoculation caused maximum attainment of lesion length within 4–6 weeks. Toxin preparation in concentration of 45 μg glucose/ml failed to evoke response.

The study demonstrates that the purified toxin being easy in storage and causing quicker response than living cells when inoculated in rice leaves, may be used to screen rice varieties for bacterial blight resistance.

Xanthomonas oryzae (UYEDA and ISHIYAMA) DOWSON causes bacterial blight of rice, a deadful disease of tropics. Potential yield loss both quantitatively and qualitatively has been encountered in rice plants due to this disease (RAO and KAUMA 1977, SINGH *et al.* 1977). Many pathogenic bacteria have been shown to produce extracellular toxins responsible for the induction of disease (STROBEL 1967, 1974, RIES and STROBEL 1972, PATIL 1974, DUBEY and SANYAL 1979), but varied opinions exist regarding the nature of the toxin produced by *X. oryzae* (EGAWA *et al.* 1968, KUO *et al.* 1970, PURUSHOTHAMAN and PRASAD 1972, RAI 1978).

As the bacterial toxins are generally proteins, polysaccharides, glycopeptides, *etc.* (STROBEL 1977) attempts have been made by various workers to grow the organism in a synthetic medium devoid of macromolecular substances, in order to purify and characterize the toxin (FINKELSTEIN *et al.* 1976, DUBEY and SANYAL 1978). Since not much attention has been paid to grow *X. oryzae* in a synthetic medium and to characterize its phytotoxin from the same, the present study was undertaken with the objectives: (i) to develop a synthetic medium for the growth of *X. oryzae*, (ii) to prepare partially purified (crude) toxin and (iii) to characterize the toxin with various temperature and pH levels.

Materials and methods

Organism: The strain of *X. oryzae* used in the present study was isolated in our laboratory from infected rice leaves, identified and maintained in potato semi-synthetic agar (WAKIMOTO 1967) stab cultures and was not allowed to undergo more than three sub-cultures.

Development of synthetic medium: To eleven 250 ml conical flasks were added NH_4NO_3 (0.029 g), K_2HPO_4 (0.125 g), KH_2PO_4 (0.125 g), Na_2SO_4 (0.025 g), $MgSO_4$ (0.0025 g), $CaCl_2$ (0.0025 g), KCl (0.0275 g), asparagine (0.025 g), sodium glutamate (0.025 g), D-glucose (0.5 g), glycerol (0.1 ml), urea (0.01 g), and micronutrient solution (0.025 ml, ELLIOT and MARGARET 1966) in such a way that each flask was devoid of one or more nutrients as shown in Table 1. Volume was made up to

Table 1
Ingredients of synthetic media and $O.D_{620}$ after 96 hr of growth of *X. oryzae*

Ingradients	Medium No.[1]										
	I	II	III	IV	V	VI	VII	VIII	IX	X	XI
Ammonium nitrate	0.029	0.029	0.150	0.029	0.029	—	—	0.029	0.029	0.029	0.029
K_2HPO_4 (g)	0.125	0.125	0.125	0.125	0.125	0.125	0.125	0.125	0.125	0.125	0.125
KH_2PO_4 (g)	0.125	0.125	0.125	0.125	0.125	0.125	0.125	0.125	0.125	0.125	0.125
Na_2SO_4 (g)	0.025	0.025	0.025	0.025	0.025	0.025	0.025	0.025	0.025	—	0.025
$MgSO_4$ (g)[2]	0.0025	0.0025	0.0025	0.0025	0.0025	0.0025	0.0025	—	0.0025	0.0025	0.0025
$CaCl_2$ (g)[2]	—	0.0025	—	0.0025	0.0025	0.0025	0.0025	0.0025	0.0025	0.0025	0.0025
KCl (g)	0.0275	—	0.0275	—	—	—	—	—	—	—	—
Asparagine (g)[2]	0.025	0.025	0.025	0.025	0.025	0.025	0.025	0.025	—	0.025	0.025
Sodium glutamate (g)[2]	—	—	—	—	—	—	—	—	0.025	—	—
D-glucose (g)[2]	0.5	0.5	0.5	—	2.0	0.5	0.5	0.5	0.5	0.5	0.5
Glycerol (ml)[2]	0.1	0.3	0.1	0.3	—	0.3	0.3	0.3	0.3	0.3	0.3
Urea (g)	0.01	0.01	—	0.01	0.01	0.01	—	0.01	0.01	0.01	0.01
Micronutrient solution (ml)	0.025	0.025	0.025	0.025	0.025	0.025	0.025	0.025	0.025	0.025	—
$O.D_{620}$	0.02	0.082	0.005	0.042	0.03	0.10	0.10	0	0.15	0.02	0.025
No. of cells/ml (x 10^7)	2	8.2	5	4.2	3.0	10	10	0	15	2	2.5

[1]) Total volume in each case was adjusted to 25 ml with deionized distilled water
[2]) Autoclaved separately and then added

25 ml with deionized distilled water and pH adjusted to 6.8. To add magnesium sulphate, calcium chloride, asparagine, sodium glutamate, D-glucose and glycerol; these ingredients were separately autoclaved and then added to respective flasks containing remaining ingredients. Flasks containing media were autoclaved at 15 lb pressure for 15 min. From pure growth of the organism in stab culture slants bacterial suspension was prepared in sterile water and optical density at 620 nm was adjusted to 0.1. Each flask after cooling was seeded with 0.1 ml of this bacterial suspension. Flasks were kept on horizontal shaker at 25 °C for continuous shaking at 120 oscillations per min for 96 hr. Optical density of cultures was observed by SPECTRONIC-20 at 620 nm and number of cells per ml was calculated. WAKIMOTO potato semisynthetic medium inoculated with *X. oryzae* in the similar way served as positive control and synthetic medium without inoculation and processed in the same way as negative control.

Time course of growth: 25 ml volumes of synthetic medium no. IX (Table 1) were added to each of 16 250 ml conical flasks. Flasks after autoclaving followed by incoulation of 0.1 ml of bacterial suspension were incubated at 25 °C for constant shaking at 120 oscillations per min. At 12 hr intervals two flasks were removed and growth was observed by measuring optical density at 620 nm in SPECTRONIC-20. The experiments were carried out in triplicate.

Preparation of culture filtrates: Culture filtrates (CF) were prepared in potato-semisynthetic media devoid of agar. In 25 ml sterile media 0.1 ml of bacterial suspension was inoculated and after 96 h of continuous shaking, contents were centrifuged in cold (4 °C) at 22,000 g for 30 min. The supernatant was dialyzed in cellophane dialysis sacks at 4 °C against potassium phosphate buffer pH 6.8, 0.04 M. The dialysate after passing through MILLIPORE membrane filter of 0.45 μm average pore diameter served as culture filtrate.

Preparation of crude toxin: This was prepared using synthetic media no. IX (Table 1). 50 ml volumes of the synthetic media in 10 500 ml conical flasks were each seeded with 0.1 ml bacterial suspension. The flasks were incubated at 25 °C for 96 h with constant shaking of 120 oscillations per min and the cultures were then centrifuged at 22,000 g for 30 min at 4 °C. The supernatants were pooled and saturated with recrystallized ammonium sulphate at 4 °C, the pH was maintained at 6.8 with ammonia solution. Suspension was held overnight at 4 °C and centrifuged at 22,000 g for 30 min at 4 °C. The precipitate was dissolved in 10 ml of 0.04 M phosphate buffer, pH 7.2, and exhaustively dialyzed against the same buffer at 4 °C. The dialysate was passed through a membrane filter (MILLIPORE) of 0.45 μm average pore diameter and stored at −20 °C. This was termed as partially purified (crude) toxin preparation.

Estimation of total carbohydrates: Total carbohydrate content of crude toxin was estimated by the method of DUBOIS *et al.* (1956) using glucose as standard.

Assay of toxicity: Toxicity of test samples was assayed by clipping the leaf tips of 30 days old rice plant cv. Taichung (Native) 1 and dipping the tips into test materials. Lesion lengths were measured upto 6 weeks of inoculation. Each test was repeated 3 times.

Time course of toxic activity: Living cell suspensions, culture filtrates and crude toxins were inoculated into rice leaves as described earlier and length of lesions developed were measured upto 6 weeks of inoculation. Living cell suspensions in sterile water served as positive control and uninoculated cut tips as negative control. Experiments were conducted in triplicate.

Effect of concentration of toxin and time of inoculation on lesion development: Serial two fold dilutions of crude toxin in sterile water were inoculated into clipped leaf tips of 30 days old T(N)1 rice plants by dipping the tips into test samples for 5, 10 and 15 seconds and lesion lengths were measured at various weeks of inoculation.

Effect of temperature on the activity of culture filtrate and crude toxin: Culture filtrate and crude toxin preparations were held at 100 °C for 10, 20 and 30 min in a water bath and heated preparations were tested for toxicity in rice leaves. Each experiment was carried out in triplicate.

Effect of pH on the activity of culture filtrate and crude toxin: Culture filtrate and crude toxin preparations were adjusted to fixed pH values from 3.0 to 10.0 in citrate phosphate buffer (pH 3.8 8.0) or ammonia solution (pH 8.0—10.0). The preparations were held at 4 °C for 12 h and then tested for toxicity at pH 6.8. Each test was done in triplicate.

Results

Growth of *X. oryzae* in various synthetic media

Table 1 shows various sets of synthetic media employed for growing of *X. oryzae*. As it is evident maximum growth with cell populations of 15 × 10^7 cells/ml was observed in medium no. IX with ingredients as ammonium nitrate and urea as nitrogen source, D-glucose and glycerol as carbon source, potassium phosphate buffer (K_2HPO_4

and KH_2PO_4), magnesium sulphate, calcium chloride and micronutrient solution. This was concluded as best suited synthetic media for the growth of *X. oryzae* and was used for further experiments. Magnesium served as an essential requirement for the growth as there was no growth in medium no. VIII which was devoid of $MgSO_4$. Asparagine retarded growth of the organism and sodium glutamate enhanced it. Glycerol and D-glucose as well as ammonium sulphate and urea when added together served as better nitrogen and carbon sources, respectively, than each added alone.

Time course of growth of *X. oryzae* in synthetic medium

Growth pattern of *X. oryzae* in synthetic media indicated almost no growth upto 12 h and thereby slow but continuous increase upto 48 h (Fig. 1) with sharp increase onwards showing maximum at 96 h with a cell population of 15.0×10^7 cells/ml.

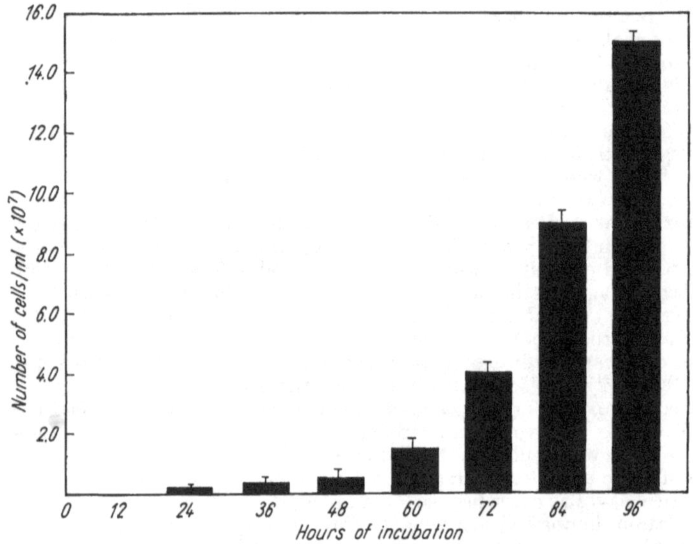

Fig. 1. Growth of *X. oryzae* in synthetic medium after various hours of incubation. Data are mean ± s.d. based on observations of three independent series

Time course of toxic activity

When culture filtrates from complex media of WAKIMOTO and crude toxins from medium no. IX were inoculated into rice leaves, typical symptoms of bacterial blight appeared within 4 days of inoculation (Fig. 2). Culture filtrate prepared from WAKIMOTO media always showed greater toxicity than crude toxin preparation. Living cells which served as positive control showed continuous lesion enlargement upto the period of 6 weeks under study. Culture filtrates and crude toxins elicited quick response in inoculated rice leaves compared to living cells and attainment of peak toxiciy in the former was also quicker, *i.e.* within 4—6 weeks than later which showed continuous increase in toxicity.

Effect of concentration of toxin and inoculation time on lesion development

When serial two fold dilutions of crude toxin preparations were inoculated in rice leaves for different time periods it was observed that lesion lengths decreased with

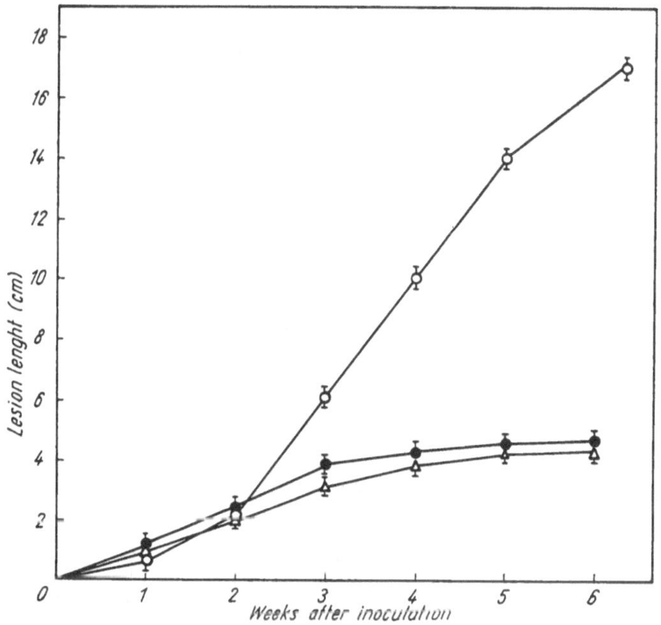

Fig. 2. Lesion lengths produced by living cells (○), culture filtrate (●) and crude toxin (△) of *X. oryzae* after various weeks of inoculation. Data are mean ± s.d. of three independent observations

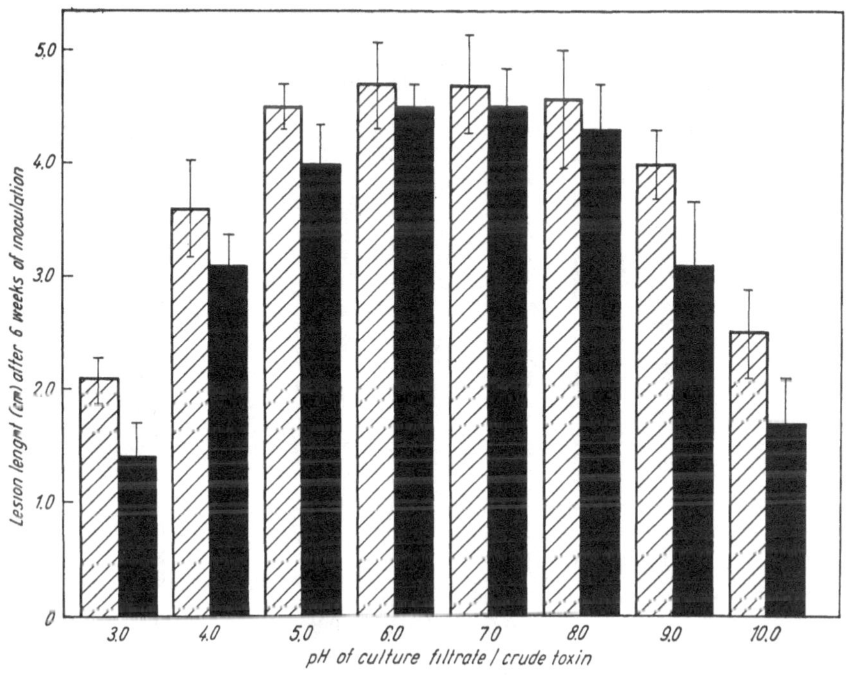

Fig. 3. Lesion lengths produced after 6 weeks of inoculation of culture filtrate (▨) and crude toxin (■) of *X. oryzae* held at different pH values. Data are mean ± s.d. of three independent observations

decrease in concentration of toxin as well as time of inoculation (Table 2). Our toxin preparation which contained 180 μg glucose/ml evoked 3.8, 4.5 and 5.0 cm lesion lengths with 5, 10 and 15 seconds of inoculation, respectively. This preparation

Table 2

Effect of toxin concentration and time of inoculation (5, 10, 15 sec) on lesion lengths (cm) after various weeks of inoculation

Weeks	Dilution of toxin								
	Undiluted			1 :2			1:4		
	5	10	15	5	10	15	5	10	15
1	0.7 ±0.2	0.9 ±0.2	1.2 ±0.3	0.2 ±0.05	0.4 ±0.4	0.5 ±0.1	0	0	0.2 ±0.1
2	1.2 ±0.3	2.0 ±0.4	3.0 ±0.2	0.5 ±0.1	0.8 ±0.2	1.2 ±0.3	0.2 ±0.1	0.3 ±0.1	0.5 ±0.2
3	1.9 ±0.3	3.1 ±0.3	3.9 ±0.5	1.0 ±0.1	1.7 ±0.2	2.0 ±0.2	0.6 ±0.1	0.8 ±0.2	1.0 ±0.2
4	2.8 ±0.5	3.8 ±0.5	4.5 ±0.4	1.7 ±0.2	2.3 ±0.3	2.6 ±0.4	0.8 ±0.1	1.1 ±0.3	1.7 ±0.2
5	3.4 ±0.5	4.2 ±0.6	4.8 ±0.6	1.9 ±0.4	2.8 ±0.2	3.0 ±0.5	0.9 ±0.2	1.4 ±0.3	2.3 ±0.2
6	3.8 ±0.6	4.5 ±0.4	5.0 ±0.9	2.0 ±0.5	3.2 ±0.6	3.5 0.4	1.6 ±0.3	1.6 ±0.1	2.5 ±0.4

Data are mean ±s.d. based on observations of three independent series

Table 3

Effect of temperature on activity of culture filtrates and crude toxins of *X. oryzae*

Weeks of inoculation	Culture filtrate/ crude toxin	Lesion length (cm)[1] after inoculation of test materials held at different temperatures			
		Unheated	100 °C 10 min	100 °C 20 min	100 °C 30 min
1	CF	1.1 ±0.3	1.0 ±0.2	0.9 ±0.2	0.6 ±0.1
	CT	0.9 ±0.2	0.9 ±0.2	0.7 ±0.1	0.4 ±0.1
2	CF	2.3 ±0.2	2.3 ±0.5	2.1 ±0.3	1.6 ±0.2
	CT	2.0 ±0.3	1.9 ±0.3	1.5 ±0.2	0.9 ±0.1
3	CF	3.8 ±0.4	3.7 ±0.3	3.4 ±0.5	2.9 ±0.3
	CT	3.1 ±0.2	3.2 ±0.5	2.7 ±0.4	2.0 ±0.2
4	CF	4.2 ±0.3	4.2 ±0.4	3.6 ±0.3	3.2 ±0.6
	CT	3.8 ±0.4	3.8 ±0.6	3.1 ±0.4	2.3 ±0.5
5	CF	4.4 ±0.6	4.3 ±0.5	3.9 ±0.4	3.4 ±0.5
	CT	4.2 ±0.5	4.0 ±0.3	3.3 ±0.6	2.5 ±0.6
6	CF	4.6 ±0.5	4.4 ±0.4	3.9 ±0.5	3.6 ±0.7
	CT	4.5 ±0.6	4.2 ±0.5	3.5 ±0.6	2.6 ±0.6

[1]) Data are mean ±s.d. of 3 replicates

in concentration of 45 μg glucose/ml failed to evoke toxic response even after 10 seconds of inoculation.

Effect of temperature on culture filtrate and crude toxin

Lesion lengths produced by culture filtrate and crude toxin preparations decreased gradually when preparations were held for increased lengths of time at 100 °C (Table 3). Almost no change in activity was noticed when preparations were held at 100 °C for 10 min, but a sharp decrease was observed when held at 100 °C for 30 min.

Effect of pH on culture filtrates and crude toxins

Culture filtrates and crude toxins of *X. oryzae* showed maximum activity when held at pH 6.0—7.0 and a gradual diminution with decreasing pH (6.0—3.0) or increasing pH (7.0—10.0). Residual activity of preparations was clearly evident even at pH 3.0 or 10.0.

Discussion

The results of the present study show the toxic properties of *X. oryzae*. Culture filtrates and partially purified toxin preparations contain a non-dialysable acid and partial heat labile glycopeptide which elicits typical blight symptoms in susceptible rice leaves similar to that caused by living cells of *X. oryzae*. The differences in lesion lengths evoked by culture filtrates and partially purified toxins were probably due to more release of the toxin in complex media.

The time course curves for lesion length caused by living cells and toxin preparations indicated continuous enlargement of lesion length in former, whereas cesession in later by 4—6 weeks. This may indicate that toxin preparations evoke maximum toxicity by 4—6 weeks beyond which there is no further toxic effect, whereas continuous multiplication of living cells leads to unchecked production of toxin causing prolonged infection and increased lesion length.

The toxic factor was more temperature and acid stable in culture filtrate than in crude toxin preparation. This might indicate more stability of the toxin in a complex medium compared to a synthetic one. Toxicity of culture filtrates and crude toxin preparations decreased drastically when they were held at 100 °C for 30 min. This rhows partial heat lability of the toxin as heat labile bacterial toxins, which are genesally proteins, loose the activity when held at 60 °C for 20 min (ANNAPURNA and SANYAL 1977) unlike heat stable toxins which are generally polysaccharides in nature and retain the activity when held at 100 °C for 30 min (SACK 1975). This suggests the nature of the toxin elaborated by *X. oryzae* to be a glycopeptide. More acid stability of this toxin compared to protein toxins confirms this, as culture filtrate and crude toxin preparations retained the activities at pH 3.0 where the protein toxins completely lose it (DUBEY and SANYAL 1979).

The present series of experiments indicate that *X. oryzae* produces a toxin which is probably implicated in the pathogenesis of blight disease of rice. Varied opinions exist about the nature of the toxin produced by *X. oryzae*. The reports are more confusing than convincing. EGAWA *et al.* (1968) suggested the nature of toxin to be phenyl acetic acid which depressed the growth of rice seedlings. PURUSHOTHAMAN and PRASAD (1972) reported a phenolic toxic compound elaborated by the organism, responsible for blight disease. According to other workers the nature of *X. oryzae* toxin was either polysaccharide (KUO *et al.* 1969) or glycopeptide (RAI 1978).

Detection and characterization of toxin(s) produced by *X. oryzae* have met with little success as in most of the cases attempts have been made to isolate the toxin from cultures of the organism in conventional complex media where constituents of the media being high molecular weight compounds interfere with purification process. To overcome such problem use of synthetic media has been advocated during recent years for purification of a wide range of bacterial toxins (Dorner 1975, Finkelstein *et al.* 1976, Dubey *et al.* 1980).

The present study suggests a synthetic medium for the growth of *X. oryzae* and the nature of the toxin elaborated in this medium as high molecular weight glycopeptide(s). Further since this toxin preparation can be stored for years together without any significant loss of activity and it elicits quicker response than living cells after inoculation in rice leaves, one set of prepared toxin can be well-employed for long periods to screen varieties for bacterial blight resistance.

Acknowledgements

The author is grateful to Dr. H. K. Pande, Director, Central Rice Research Institute, Cuttack for providing facilities and encouragement and to Dr. S. B. Lodh, Head, Division of Biochemistry, for critical comments an suggestions.

References

Annapurna, E. and Sanyal, S. C., 1977. Enterotoxicity of *Aeromonas hydrophila*. J. Med. Microbiol., **10**, 317—323.

Dorner, F., 1975. *Escherichia coli* enterotoxin: Purification and partial characterisation. J. biol. Chem., **250**, 8712—8719.

Dubey, R. S. and Sanyal, S. C., 1978. Enterotoxicity of *Aeromonas hydrophila*: Skin responses and *in vivo* neutralisation. Zbl. Bakt. Hyg., I. Abt. Orig. A, **242**, 487—499.

Dubey, R. S. and Sanyal, S. C., 1979. Studies on the characterisation and neutralisation of *Aeromonas hydrophila* enterotoxin in the rabbit ileal loop model. J. Med. Microbiol., **12**, 345 to 352.

Dubey, R. S., Sanyal, S. C. and Malhotra, O. P., 1980. Purification of *Aeromonas hydrolphila* enterotoxin and its mode of action in experimental model. In: Natural Toxins (Eds. D. Eaker and T. Wadstrom). Pergamon Press Oxford and New York, 259—268.

Dubois, M., Gilles, K. A., Hamilton, J. K., Rebers, P. A. and Smith, F., 1956. Colorimetric method for determination of sugars and related substances. Anal. Chem., **28**, 350—356.

Egawa, H., Yoshii, K. and Ueyama, A., 1968. Phenylacetic acid, a metabolite of *Xanthomonas oryzae* (Uyeda and Ishiyama) Dowson in culture affecting upon the derepressive growth of young roots of rice seedlings. Ann. Phytopath. Soc., Japan, **34**, 46—50.

Elliot, C. G. and Margaret, R. H., 1966. The sterol requirement of *Phytophthora cactorum*. J. gen. Microbiol., **42**, 425—435.

Finkelstein, R. A., Larue, M. K., Johnston, D. W., Vasil, M. L., Cho, G. J. and Jones, J. R., 1976. Isolation and properties of heat labile enterotoxin(s) from enterotoxigenic *E. coli*. J. Infect. Dis., **133**, S 120—137.

Kuo, T. T., Ling, B. C. and Li, C. C., 1970. Bacterial leaf blight of rice plant. III. Phytotoxic polysaccharides produced by *Xanthomonas oryzae*. Bot. Bull. Acad. Sinica, **11**, 46—54.

Patil, S. S., 1974. Toxins produced by phytopathogenic bacteria. Ann. Rev. Phytopathol., **12**, 259—278.

Purusothaman, D. and Prasad, N. N., 1972. Isolation of a toxin from *Xanthomonas oryzae*. Phytopath. Z., **75**, 175—180.

Rai, P. V., 1978. Toxins produced by a virulent and a weakly virulent strain of *Xanthomonas oryzae*. Indian Phytopath., **31**, 60—64.

Rao, P. S. and Kauma, H. E., 1977. Potential yield losses in dwarf rice varieties due to bacterial blight in India. Phytopath. Z., **90**, 281—284.

Ries, S. M. and Strobel, G. A., 1972. Biological properties and pathological role of a phytotoxic glycopeptide from *Corynebacterium insidiosum*. Physiol. Plant Path., **2**, 133—142.

Sack, R. B., 1975. Human diarrhoeal disease caused by enterotoxigenic *Escherichia coli*. Ann. Rev. Microbiol., **29**, 333—353.

SINGH, G. P., SRIVASTAVA, M. K., SINGH, R. V. and SINGH, R. M., 1977. Variation in quantitative and qualitative losses caused by bacterial blight in different rice varieties. Indian Phytopath., **30**, 180—185.

STROBEL, G. A., 1967. Purification and properties of phytotoxic polysaccharide produced by *Corynebacterium sepedonicum*. Plant Physiol., **42**, 1433—1441.

STROBEL, G. A., 1974. Phytotoxins produced by plant parasites. Ann. Rev. Plant. Physiol., **25**, 541—566.

WAKIMOTO, S., 1967. Strains of *Xanthomonas oryzae* in Asia and their virulence against rice varieties. In: Symposium on Rice Diseases and their Control by Growing Resistant Varieties and Other Measures. Ministry of Agriculture and Forestry, Japan, September 25—28, B 1 11.

Mailing address: Dr. R. S. DUBEY
Biochemistry Section
Department of Chemistry
Banaras Hindu University
Varanasi-221005, India

Zeitschrift für Allgemeine Mikrobiologie	23	1983	4	235–246

(Zentralinstitut für Mikrobiologie und experimentelle Therapie
der Akademie der Wissenschaften der DDR und
Sektion Biologie der Friedrich-Schiller-Universität Jena)

Degradation of aniline and monochloroanilines by *Rhodococcus* sp. An 117 and a pseudomonad: A comparative study

U. KAMINSKI, D. JANKE, H. PRAUSER and W. FRITSCHE

(Eingegangen am 16. 8. 82)

Two newly isolated aniline-degrading bacterial strains were characterized with regard to their enzyme systems responsible for aniline catabolism. One of them identified as a *Rhodococcus* sp. metabolized aniline exclusively via the β-ketoadipate pathway by means of inducible enzymes. The aniline-degrading enzyme system of the second isolate, presumably a pseudomonad, was shown to consist of an inducible aniline-converting enzyme and constitutive meta-pathway enzymes. Both isolates failed to metabolize monochlorinated anilines in the absence of additional carbon sources. To explain this the ring-cleaving enzymes of both isolates were examined for their substrate specificities. Furthermore, the effect of 4-chlorocatechol on the enzymes catalyzing aniline conversion and catechol oxygenation was investigated.

Microbial degradation of phenyl amides is of environmental relevance. Thus aniline and its derivatives can be found among the waste products of certain chemical industries. In agricultural soils treated with herbicides of the phenyl amide type (phenylureas, phenylcarbamates, acylanilides), chloroanilines are formed by microbial transformation of the man-made chemicals. In some cases the chloroanilines which may react to produce hazardous azocompounds (BARTHA and PRAMER 1970) when accumulated in soil proved to be even more persistent than the applied herbicides.

Soil bacteria capable of utilizing aniline as sole source of carbon, nitrogen, and energy normally fail to degrade chlorinated anilines. However, some of them do metabolize these compounds in the presence of additional energy-providing carbon sources (REBER *et al.* 1979, SUROVTSEVA *et al.* 1980, YOU and BARTHA 1982).

This paper presents the first results of an investigation concerning microbial degradation of chlorinated anilines. It deals mainly with the characterization of two newly isolated aniline-degrading bacteria and their enzyme systems responsible for aniline catabolism.

Materials and methods

Isolation of the bacterial strains and their identification are described in the experimental part.

Cultivation: Both strains An 117 and SB 3 were maintained at 4 °C on agar slants using the R (rich)-medium of KAZUHIKO and KOMAGATA (1972). Liquid cultures were grown at 28 °C in a mineral salts medium containing per litre: 9.0 g $Na_2HPO_4 \cdot 12\ H_2O$, 1.5 g KH_2PO_4, 0.2 g $MgSO_4 \cdot 7\ H_2O$, 0.01 g $CaCl_2$, 0.0025 g $FeSO_4 \cdot 7\ H_2O$, 0.5 mg H_3BO_4, 0.1 mg KJ, 0.4 mg $MnSO_4 \cdot 4\ H_2O$, 0.4 mg $ZnSO_4 \cdot 7\ H_2O$, 0.2 mg $Na_2MoO_4 \cdot 2\ H_2O$, 0.1 mg $CuSO_4 \cdot 5\ H_2O$; pH was 6.9. When carbon sources other than aniline were used as substrates, the medium was supplemented with 1 g/l NH_4NO_3. Nature and concentration of the carbon sources used are mentioned in the experimental part.

Cell-free extracts: Crude extracts were prepared as described by SALA-TREPAT and EVANS (1971) without the high-speed centrifugation step. The protein was estimated by the method of LOWRY *et al.* (1951) using bovine serum albumin as standard.

Enzyme assays: The activity of catechol 2,3-dioxygenase (metapyrocatechase, EC 1.13.1.2) was determined according to NOZAKI (1970). The activity of the enzymes metabolizing 2-hydroxymuconic semialdehyde (HMS), 2-hydroxy-5-methylmuconic semialdehyde (HMMS) and 2-hydroxy-5-chloromuconic semialdehyde (HClMS) was determined by measuring the rate of decrease in absorbance at 375, 382 and 379 nm, respectively. NAD^+ (0.30 μmoles) was added to the reaction mixture where indicated. The semialdehydes were obtained as described by SALA-TREPAT and EVANS (1971) by using heat-treated extracts from strain SB 3 cells. All calculations were made on the basis of the following molar extinction coefficients: HMS-44000 (NOZAKI 1970), HMMS-28100 (BAYLY *et al.* 1966), HClMS-39600 (SALA-TREPAT and EVANS 1971). The activity of catechol 1,2-dioxygenase (EC 1.13.1.1) was determined according to NAKAZAWA and NAKAZAWA (1970) by measuring product formation at 260 nm. The respective specific activities were calculated on the basis of the molar extinction coefficients given by DORN and KNACKMUSS (1978).

All enzyme assays were performed at room temperature in a SPECORD recording spectrophotometer. The unit of enzyme activity is defined as the amount of enzyme necessary to convert 1 μmole of substrate to the respective product in 1 min under the conditions used.

Whole-cell activities: These were determined by using resting cells washed and resuspended in 0.1 M phosphate buffer, pH 6.9, containing chloramphenicol (150 μg/ml). Manometric experiments were performed as described previously (JANKE and FRITSCHE 1979).

Biomass concentration: The optical density of the culture medium was measured at 578 nm in a SPEKOL-spectrophotometer and the dry weight of cells was determined by using a calibration curve.

Colourimetric methods: Aniline was determined by the method of BRATTON *et al.* (1939) as modified by RABSCH and FRITSCHE (1977). 4-Chlorocatechol was estimated according to ARNOW (1937).

Electron microscopy: Specimens for electron microscopy were obtained by fixing cell suspensions with osmiumtetroxide and shadowing (30°) with a platinum-carbon mixture or negative staining according to the method of BRENNER and HORNE (1959).

Results

Isolation and characterization of aniline-degrading bacteria (Table 1)

The aniline-degrading bacterial strain An 117 (Figs. 1, 2) was directly isolated from a chemostat culture in which it outgrew aniline-fed cells of *Achromobacter* sp. Ir 2. Most likely this contaminant originated from enrichment cultures performed simultaneously. Morphologically strain An 117 resembled in some respects the nocardia-like aniline-utilizing soil bacterium described by BACHOFER *et al.* (1973). Phage-typing with a set of phages active against the *Nocardia/Rhodococcus* complex (PRAUSER 1981) clearly indicated that strain An 117 belonged to either one of these genera. Cell and colony morphologies were typical of the genus *Rhodococcus* (ZOPF) GOODFELLOW and ALDERSON 1977.

Table 1
Properties of strains An 117 and SB 3

	An 117	SB 3
colony morphology[a]	round, smooth convex, opaque	round, smooth convex, opaque
colony colour[a]	coral	colourless
cell morphology[a]	nocardioform	rods
motility	non-motile	motile
GRAM-reaction	GRAM-positive	GRAM-negative
rormation of spores	no	no
anaerobic acid production[b]	no	no
arginine dihydrolase[c]	absent	present
resistance towards 0.1% 2, 3, 5-triphenyltetrazolium chloride	not tested	resistant

[a]) after incubation at 28 °C for 2 days on R-medium agar
[b]) tested according to HUGH and LEIFSON (1953)
[c]) tested according to THORNLEY (1960)

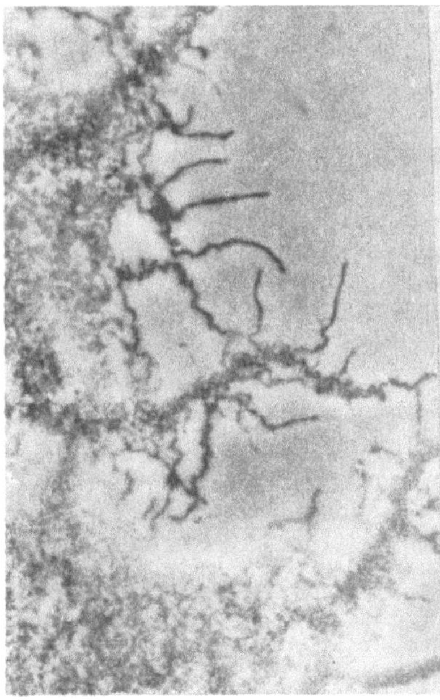

1

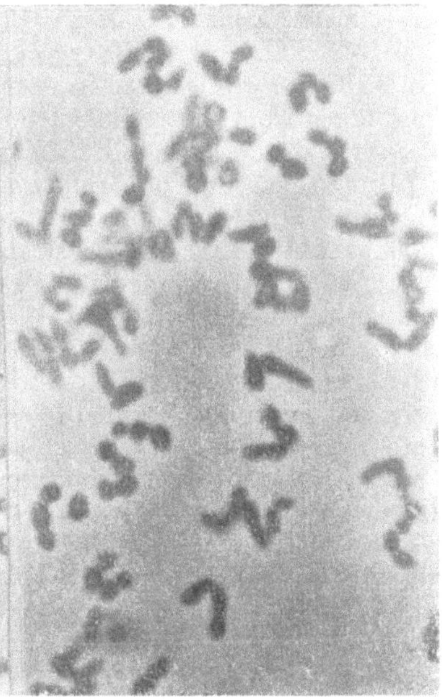

2

Fig. 1. *Rhodococcus* sp. An 117. Fragmented mycelium on water agar, 4 days, 28 °C. Magnification: 400:1

Fig. 2. *Rhodococcus* sp. An 117. Mycelial fragments formed in agitated cultures grown in dextrose meat extract medium for 48 hours at 28 °C. Magnification: 2000:1

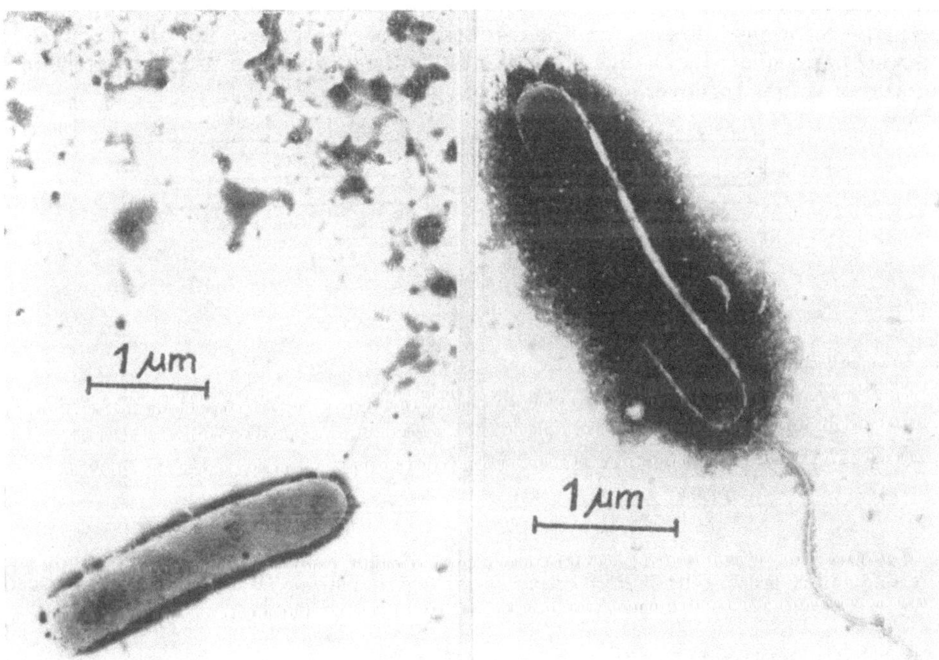

3a 3b

Fig. 3. Electron micrographs of strain SB 3 cells.
a) negatively stained with potassium phosphotungstate
b) shadowed (30°) with a platinum-carbon mixture

Strain SB 3 (Fig. 3) was isolated from conventional enrichment cultures established with samples taken from the activated-sludge system of a chemical plant (VEB Synthesewerk Schwarzheide). 100-ml-portions of the mineral salts medium containing aniline (50 μl/l) as sole source of carbon and nitrogen were inoculated with 5-ml-samples of the activated sludge and incubated in 500-ml-flasks on a rotary shaker at 28 °C. After 2 days at which time the substrate had disappeared completely two serial transfers were made to fresh media (aniline concentration, 200 μl/l). From the third passage (aniline concentration, 300 μl/l) diluted culture medium was plated on mineral salts agar medium containing aniline (500 μl/l) as sole source of carbon and nitrogen. After incubation at 28 °C for 7 days numerous uniform colonies developed from which strain SB 3 was obtained by repeated single-colony isolation.

Cells of strain SB 3 represented Gram-negative rods motile by means of single polar flagellas as demonstrated by electron microscopy. No peritrichously flagellated cells were observed. This organism seems to be a pseudomonad since some additional properties (resistance towards 0.1% 2, 3, 5-triphenyltetrazolium chloride; presence of arginine dihydrolase) were typical of members of the genus *Pseudomonas*.

Characterization of the aniline-degrading enzyme systems

Both isolates were able to utilize aniline as sole source of carbon and nitrogen whereas chloroanilines could not be used as growth substrates. In order to test whether the latter compounds were degraded without being assimilated, we performed experiments with resting cells grown on aniline. As shown in Fig. 4 both strains An 117 and SB 3 converted aniline with a high rate (0.7 and 0.4 μmoles $\cdot$ h^{-1} $\cdot$ mg^{-1} dry weight of cells, respectively) but failed to metabolize monochloroanilines. To elucidate the pathway as well as the type of regulation of aniline degradation in the two strains the enzymatic reactions initiating aniline degradation were studied. For this purpose the cells were preincubated with different potential inducers and than used for determining the rates of aniline turnover and catechol oxygenation.

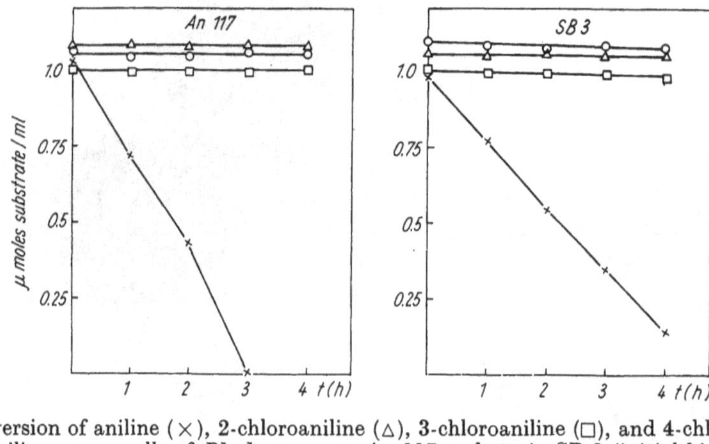

Fig. 4. Conversion of aniline (×), 2-chloroaniline (△), 3-chloroaniline (□), and 4-chloroaniline (□) by resting aniline-grown cells of *Rhodococcus* sp. An 117 and strain SB 3 (initial biomass concentrations were 0.5 mg dry weight of cells/ml)

Rhodococcus sp. An 117

The results summarized in Table 2 demonstrate that in *Rhodococcus* sp. An 117 both reactions studied were catalyzed by inducible enzymes. Aniline as well as its monochlorinated derivatives proved to be capable of inducing (or derepressing) the synthesis of the enzymes responsible for aniline convertion and ortho-cleavage of catechol. Cells preincubated with catechol exhibited only the latter activity. No meta-dyrocatechase was found in any of the different cell preparations indicating that aniline is metabolized exclusively via the β-ketoadipate pathway.

Table 2
Activities of the aniline-degrading enzyme system in *Rhodococcus* sp. An 117 cells preincubated with different compounds (figures in parentheses indicate the relative activities in %)

substrate used for preincubation[a])	mM	growth during preincubation	aniline turnover rate[b])	specific activity of catechol 1,2-dioxygenase	specific activity of catechol 2,3-dioxygenase
succinate	13	growth	0.03 (4)	0.016 (6)	$<$ 0.001
aniline	5	growth	0.67 (100)	0.261 (100)	$<$ 0.001
2-chloroaniline	1	no growth	0.24 (36)	0.498 (191)	$<$ 0.001
3-chloroaniline	1	no growth	0.30 (45)	0.236 (90)	$<$ 0.001
4-chloroaniline	1	no growth	0.11 (16)	0.238 (91)	$<$ 0.001
catechol	5	growth	0.04 (6)	0.225 (86)	$<$ 0.001

[a]) incubation period was 4 hours with catechol and 16 hours with the remaining substrates
[b]) rates (expressed as μmoles of aniline converted $\times$ h^{-1} $\times$ mg^{-1} dry weight of cells) were determined by following the conversion of aniline (initial conc. 1 mM) within 3 to 4.5 hours by whole cells washed and resuspended in 0.1 M phosphate buffer, pH 6.9, containing chloramphenicol (150 μg/ml)

Strain SB 3

Strain SB 3 behaved differently from An 117. In crude cell extracts no catechol 1,2-dioxygenase activity was detectable (Table 3) whereas surprisingly high levels of metapyrocatechase were found even in cells precultivated with succinate. This ring-cleaving activity could be observed also in cells maintained for several weeks on R-medium agar slants. Conversion of aniline, however, required cell precultivation with aniline as demonstrated by manometric experiments (Table 4). Meta-

Table 3
Activities of the catechol dioxygenases and the hydroxymuconic semialdehydes metabolizing enzymes in strain SB 3 cells (figures in parentheses indicate the relative activities in %)

substrate used for cell precultivation	mM	specific activity of catechol 1,2-dioxygenase	specific activity of catechol 2,3-dioxygenase	specific activity of the HMS-metabolizing enzymes[a])	
				without NAD$^+$	with NAD$^+$
succinate	13	$<$ 0.001	3.351 (58)	0.014 (56)	0.057 (52)
aniline	2.5	$<$ 0.001	5.780 (100)	0.025 (100)	0.109 (100)

[a]) HMS solution was added to the reaction mixture to give an initial absorbance at 375 nm of 1.0 to 1.2

Table 4
Relative rates of oxygen uptake (Q_{O_2}) during aniline oxidation by resting cells of strain SB 3

substrate used for cell precultivation (mM)	Q_{O_2}[a])	μmoles O_2 / μmole aniline
succinate (13)	4	—
aniline (2.5)	106	5.15

[a]) expressed in μl O_2 $\times$ h^{-1} $\times$ mg^{-1} dry weigth of cells; the values were corrected for endogenous respiration

pyrocatechase activity proved to be closely correlated with the level of the enzymes metabolizing hydroxymuconic semialdehydes. The latter was significantly higher when NAD^+ (0.30 μmoles) was added to the reaction mixture. This indicates the presence of a semialdehyde dehydrogenase. But even in the presence of the cofactor the specific activity was about 50 times lower in comparison with that of metapyrocatechase (Table 3).

Substrate specificity of the ring-cleaving enzymes

Certain bacterial strains are known to synthesize enzymes specific for degradation of chlorinated aromatic compounds (BOLLAG *et al.* 1968, GAUNT and EVANS 1971, DORN and KNACKMUSS 1978, CRAWFORD *et al.* 1979). To clarify whether this is also true of *Rhodococcus* sp. An 117 and strain SB 3 and, in addition, to obtain further information on their degradative potentials, both strains were incubated with aniline or one of its monochlorinated derivatives and the respective catechol-oxygenating enzymes synthesized were examined for their substrate specificities.

Rhodococcus sp. An 117

The experiments with *Rhodococcus* sp. An 117 revealed a rather constant pattern of substrate specificity of catechol 1,2-dioxygenase in cell-free extracts irrespective of whether aniline, monochloroanilines or catechol were used for cell preincubation (Table 5). In all cases the ring-cleaving activity towards 4-methylcatechol was 50 to 60% of that observed with catechol, whereas 4-chlorocatechol was oxygenated with a rate 17 times lower than that seen with the unsubstituted analogue.

Table 5
Relative activities (in %) of catechol 1,2-dioxygenase towards 4-methylcatechol (4 MCat) and 4-chlorocatechol (4 ClCat) in cells of *Rhodococcus* sp. An 117 preincubated with different aromatic compounds

substrate used for preincubation (mM)[a]	relative activities [b]) of catechol 1,2-oxygenase towards		
	catechol	4MCat	4ClCat[c])
aniline (5)	100	61	7
2-chloroaniline (1)	100	67	5
3-chloroaniline (1)	100	48	6
4-chloroaniline (1)	100	66	6
catechol (5)	100	58	5

[a]) incubation period was 16 hours except for catechol (4 hours)
[b]) the respective specific activities towards catechol are given in Table 2
[c]) initial substrate concentration in the reaction mixture, 0.34 mM

Strain SB 3

The meta-pathway enzymes in strain SB 3 cells proved to be rather unspecific. Thus crude extracts from succinate-grown cells oxygenated 4-chlorocatechol as rapidly as catechol whereas 4-methylcatechol was cleaved with a higher rate than catechol (Table 6). A similar ratio of meta-cleaving activity towards catechol and its two derivatives was observed with extracts prepared from aniline-grown cells, suggesting that a single non-specific metapyrocatechase is involved in aniline degradation by strain SB 3 cells.

The substituted hydroxymuconic semialdehydes resulting from meta-cleavage of 4-chlorocatechol and 4-methylcatechol were metabolized with rates of 30 and 170%, respectively, as compared with the 2-hydroxymuconic semialdehyde regardless of whether extracts from succinate-grown or aniline-grown cells were used (Table 7). Thus evidence was obtained that during growth on aniline strain SB 3 synthesizes no hydroxymuconic semialdehydes metabolizing enzymes other than those found after cell precultivation with non-aromatic compounds.

Table 6
Relative activities (in %) of catechol 2,3-dioxygenase towards 4-methylcatechol (4MCat) and 4-chlorocatechol (4ClCat) in strain SB 3 cells

substrate used for cell pre-cultivation (mM)	relative activities[a]) of catechol 2,3-dioxygenase towards		
	catechol	4MCat	4ClCat[b])
succinate (13)	100	156	100[c])
aniline (2.5)	100	157	107[c])

[a]) the respective specific activities towards catechol are given in Table 3
[b]) initial substrate concentration in the reaction mixture, 0.34 mM
[c]) enzyme inactivation was observed after initial linear product formation (see also Fig. 7)

Table 7
Relative activities (in %) of the hydroxymuconic semialdehyde dehydrogenase in strain SB 3 cells

substrate used for cell pre-cultivation (mM)	relative activities[a]) towards		
	HMS	HMMS	HClMS
succinate (13)	100	177	31
aniline (2.5)	100	169	29

[a]) the respective specific activities towards HMS are given in Table 3; all activities were determined in the presence of NAD^+ (0.30 μmoles); semialdehyde solutions were added to the reaction mixtures to give an initial absorbance of 1.0 to 1.2 at the respective wave length

Effect of 4-chlorocatechol on aniline turnover and catechol oxygenation

The following experiments were aimed at studying the effect of 4-chlorocatechol on the enzymes initiating aniline degradation in *Rhodococcus* sp. An 117 and strain SB 3. Chlorinated catechols are well-established metabolites in microbial degradation of chloroanilines. Recently evidence has been obtained which shows that these intermediates can directly affect the velocity of degradative processes by inhibiting key enzymes of the bacterial catabolism of aromatic compounds (REBER and THIERBACH 1980, SUROVTSEVA and VOLNOVA 1981, KLECKA and GIBSON 1981).

Whole resting aniline-grown cells of the strains An 117 and SB 3 revealed a strong inhibition of aniline turnover by 4-chlorocatechol (Fig. 5). The extent of inhibition depended on the initial 4-chlorocatechol concentration and was rather similar for the two strains. This was true even though strain SB 3 cells completely converted the inhibitor into the corresponding hydroxymuconic semialdehyde within a short period of time, due to the very high level of metapyrocatechase. To demonstrate the effect of 4-chlorocatechol on the ring-cleaving enzymes, several enzymatic experiments were performed. The results obtained with crude extracts from aniline-grown cells of *Rhodococcus* sp. An 117 suggest that the chlorinated aromatic compound acts as a competitive inhibitor of the respective catechol 1,2-dioxygenase (Fig. 6). It also affected metapyrocatechase activity in crude extracts from strain SB 3 cells. As shown in Fig. 7a the amount of chlorinated hydroxymuconic semialdehyde formed from 4-

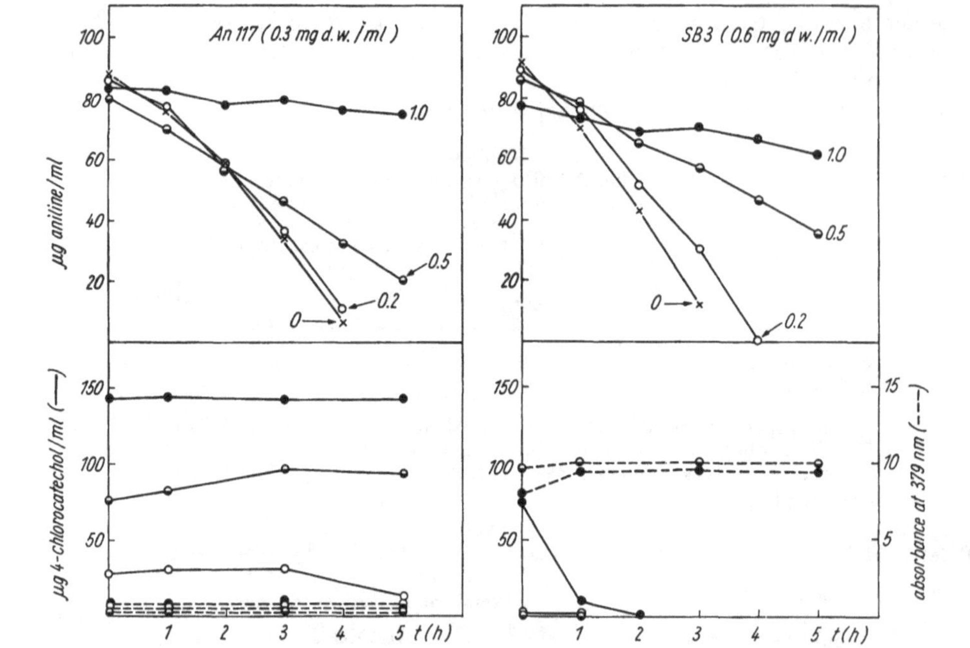

Fig. 5. Effect of 4-chlorocatechol on aniline conversion by resting aniline-grown cells of *Rhodococcus* sp. An 117 and strain SB 3 (parameters indicate the initial concentration, expressed in mM of the exogenously supplied 4-chlorocatechol; cells were preincubated with 4-chlorocatechol for 10 min before aniline was added)

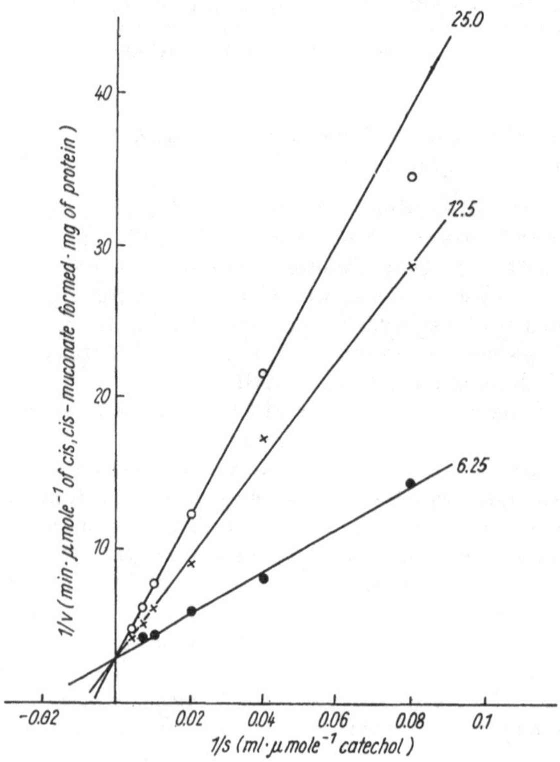

Fig. 6. Effect of 4-chlorocatechol on the activity of catechol 1,2-dioxygenase in crude extracts from aniline-grown cells of *Rhodococcus* sp. An 117 towards catechol (parameters correspond to μmoles of 4-chlorocatechol per ml reaction mixture)

chlorocatechol as well as the rate of this conversion decreased remarkably with increasing initial concentration of 4-chlorocatechol. Furthermore, the amount of chlorinated hydroxymuconic semialdehyde formed was found to depend directly upon the amount of enzyme added to the reaction mixture, indicating a rapid inactivation of metapyrocatechase by the chlorinated compound (Fig. 7b).

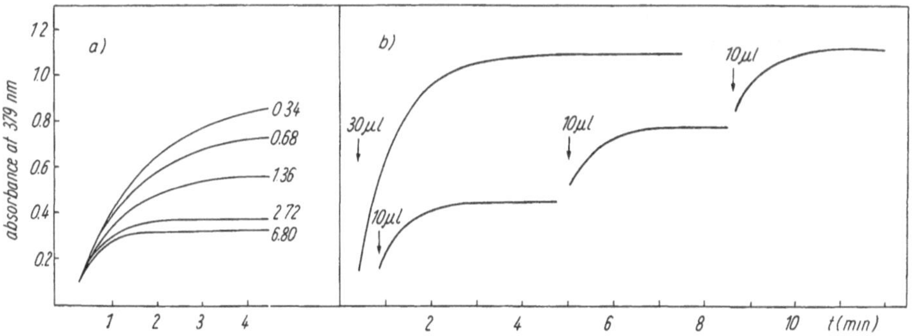

Fig. 7. Kinetics of 4-chlorocatechol conversion by crude extracts from strain SB 3 cells:
a) Effect of substrate concentration (protein content in the reaction mixture, 2.05 µg; parameters correspond to µmoles of 4-chlorocatechol per ml reaction mixture)
b) Relationship between the amount of crude extract (µl) added to the reaction mixture and the amount of 4-chlorocatechol converted (initial substrate concentration, 2.72 mM)

Discussion

Phenyl amides, in contrast to many other aromatic compounds, are rarely found in nature. Therefore it is not surprising that only a small number of aniline-degrading soil bacteria could be isolated. Twelve different isolates capable of utilizing aniline as the sole source of carbon, nitrogen, and energy have been described in the literature. They include strains of *Alcaligenes faecalis* (WALKER anf HARRIS 1969, SUROVTSEVA and VOLNOVA 1980), soil bacteria of the *Arthrobacter* and *Achromobacter* groups (CLARK and WRIGHT 1970, RABSCH and FRITSCHE 1977), pseudomonads (HELM and REBER 1979, YOU and BARTHA 1982) as well as nocardioform organisms (v. PLOTHO 1950, MOORE 1949, BACHOFER *et al.* 1973). The present work extends the range of bacterial strains characterized with regard to their aniline-degrading enzyme systems.

Strain An 117 identified as a *Rhodococcus* sp. was found to metabolize aniline exclusively via the β-ketoadipate pathway by means of inducible enzymes. Synthesis of its catechol 1,2-dioxygenase is regulated independently of the aniline-converting enzyme. Thus in cells incubated with catechol high levels of the ring-cleaving enzyme could be demonstrated whereas the aniline converting activity was very low. This situation is consistent with catechol 1,2-dioxygenase synthesis being induced by cic, cis-muconic acid, as it was reported previously for several bacterial ortho-cleavage systems (ORNSTON 1966, CANOVAS and STANIER 1967, BIRD and CAIN 1968, RANN and CAIN 1969).

Catechol 1,2-dioxygenase of *Rhodococcus* sp. An 117 exhibited only low activity towards 4-chlorocatechol which is in agreement with the theoretical considerations of KNACKMUSS (1975) according to which para-halogen substituents retard the electrophilic attack of „ordinary" pyrocatechases. On the other hand, 4-chlorocatechol was found to inhibit competitively the activity of the enzyme in *Rhodococcus* sp. An 117 towards catechol.

The second isolate studied — strain SB 3 — was presumably a pseudomonad. Its aniline-degrading enzyme system differed considerably from that of *Rhodococcus* sp. An 117. Thus aniline catabolism in SB 3 cells is initiated by the action of an inducible enzyme. The subsequent breakdown of the resulting catechol, however, is mediated by constitutive catechol 2,3-dioxygenase and hydroxymuconic semialdehyde dehydrogenase. Constitutive metapyrocatechases were also found in certain strains of *Pseudomonas putida* (FEIST and HEGEMAN 1969). Nevertheless, this type of regulation is generally regarded as uncommon for peripheric catabolic enzymes in natural isolates. Thus constitutivity of meta-pathway enzymes in strain SB 3 might also be related to the selective conditions specific of the habitat from which the organism was isolated. A single catechol 2,3-dioxygenase seems to catalyze catechol oxygenation in strain SB 3. This enzyme proved to be rather unspecific because it was highly active towards 4-methylcatechol and 4-chlorocatechol as well. The latter compound, however, rapidly inactivated the enzyme when present in sufficiently high concentrations.

Catechol 2,3-dioxygenase and hydroxymuconic semialdehyde dehydrogenase are apparently regulated coordinately in strain SB 3 since their levels were found to be closely correlated irrespectively of whether succinate or aniline was used as growth substrate. The ratio of their specific activities, however, was about 50:1. A similar striking discrepancy between the levels of metapyrocatechase and hydroxymuconic semialdehyde dehydrogenase was reported for cells of *Pseudomonas* P_G grown on naphthalene or 2-methylnaphthalene (WILLIAMS *et al.* 1975).

Like other aniline-degrading bacteria (WALKER and HARRIS 1969, REBER *et al.* 1979, SUROVTSEVA *et al.* 1980), both *Rhodococcus* sp. An 117 and strain SB 3 failed to metabolize any of the monochloroanilines in the absence of additional carbon sources. In cells of the former isolate, however, each of the monochloroanilines proved to be able to induce significantly the enzymes initiating aniline catabolism. This indicates that the failure of *Rhodococcus* sp. An 117 to metabolize the chlorinated compounds has reasons other than insufficient penetration of the monochloroanilines into the cells. Among those worth considering are: low or missing activity of the aniline-oxygenating enzyme towards chlorinated anilines, toxicity (or inhibitory effects) of the latter compounds and/or their metabolites, insufficient provision of reduced pyridine nucleotides to the aniline-oxygenating enzyme. To decide between these possibilities we are presently studying the effect of certain energy-providing carbon sources on the degradation of monochloroanilines by *Rhodococcus* sp. An 117.

Acknowledgements

We thank our colleagues in the laboratory for electron microscopy of the Central Institute of Microbiology and Experimental Therapy for their technical assistance.

References

ARNOW, L. E., 1937. Colorimetric determination of the components of 3,4-hydroxyphenylalanine-tyrosine mixtures. J. biol. Chem., **118**, 531.

BACHOFER, R., OLTMANNS, O. and LINGENS, F., 1973. Isolation and characterization of a *Nocardia*-like soil bacterium growing on carboxyanilide fungicides. Arch. Mikrobiol., **90**, 141.

BARTHA, R. and PRAMER, D., 1970. Metabolism of acylanilide herbicides. Adv. Appl. Microbiol., **13**, 317.

BAYLY, R. C., DAGLEY, S. and GIBSON, D. T., 1966. The metabolism of cresols by species of *Pseudomonas*. Biochem. J., **101**, 293.

BIRD, J. A. and CAIN, R. B., 1968. Cis, cis-muconate, the product inducer of catechol 1,2-oxygenase in *Pseudomonas aeruginosa*. Biochem. J., **109**, 479.

BOLLAG, J.-M., BRIGGS, G. G., DAWSON, J. E. and ALEXANDER, M., 1968. Enzymatic degradation of chlorocatechols. J. agric. Food Chem., **16**, 829.

BRATTON, A. C., MARSHALL, E. K., Jr., BABBITT, D. and HENDRICKSON, A. R., 1939. A new coupling component for sulfanilamide determination. J. biol. Chemistry, **128**, 537.

BRENNER, S. and HORNE, R. W., 1959. A negative staining method for high resolution electron microscopy of viruses. Biochim. biophysica Acta, **34**, 103.

CANOVAS, J. L. and STANIER, R. Y., 1967. Regulation of the enzymes of the β-ketoadipate pathway in *Moraxella calcoacetica*. I. General aspects. Eur. J. Biochem., **1**, 289.

CLARK, C. G. and WRIGHT, S. J. L., 1970. Detoxication of isopropyl N-phenylcarbamate (IPC) and isopropyl N-3-chlorophenylcarbamate (CIPC) in soil, and isolation of IPC-metabolizing bacteria. Soil Biol. Biochem., **2**, 19.

CRAWFORD, R. L., OLSON, P. E. and FRICK, T. D., 1979. Catabolism of 5-chlorosalicylate by a *Bacillus* isolated from the Mississippi River. Appl. environ. Microbiol., **38**, 379.

DORN, E. and KNACKMUSS, H.-J., 1978. Chemical structure and biodegradability of halogenated aromatic compounds. Substituent effects on 1,2-dioxygenation of catechol. Biochem. J., **174**, 85.

FEIST, C. F. and HEGEMAN, G. D., 1969. Phenol and benzoate metabolism by *Pseudomonas putida*: regulation of tangential pathways. J. Bacteriol., **100**, 869.

GAUNT, J. K. and EVANS, W. C., 1971. Metabolism of 4-chloro-2-methylphenoxyacetate by a soil pseudomonad. Ring fission, lactonizing and delactonizing enzymes. Biochem. J., **122**, 533.

HELM, V. and REBER, H., 1979. Investigation on the regulation of aniline utilization in *Pseudomonas multivorans* strain An-1. Eur. J. appl. Microbiol. Biotechnol., **7**, 191.

HUGH, R. and LEIFSON, E., 1953. The taxonomic significance of fermentative versus oxidative metabolism of carbohydrates by various gram-negative bacteria. J. Bacteriol., **66**, 24.

JANKE, D. und FRITSCHE, W., 1979. Dechlorierung von 4-Chlorphenol nach extradioler Ringspaltung durch *Pseudomonas putida*. Z. Allg. Mikrobiol., **19**, 139.

KAZUHIKO, Y. and KOMAGATA, K., 1972. Taxonomic studies on coryneform bacteria. IV. Morphological, cultural, biochemical, and physiological characteristics. J. gen. appl. Microbiol., **18**, 399.

KLECKA, G. M. and GIBSON, D. T., 1981. Inhibition of catechol 2,3-dioxygenase from *Pseudomonas putida* by 3-chlorocatechol. Appl. environ. Microbiol., **41**, 1159.

KNACKMUSS, H.-J., 1975. Über den Mechanismus der biologischen Persistenz von halogenierten raomatischen Kohlenwasserstoffen. Chem.-Ztg., **99**, 213.

LOWRY, O. H., ROSEBROUGH, N. J., FARR, A. L. and RANDALL, R. J., 1951. Protein measurement with the Folin phenol reagent. J. biol. Chemistry, **193**, 265.

MOORE, F. W., 1949. The utilization of pyridine by microorganisms. J. gen. Microbiol., **3**, 143.

NAKAZAWA, T. and NAKAZAWA, A., 1970. Pyrocatechase (*Pseudomonas*). In: Methods in Enzymology, **17a**, 518.

NOZAKI, M., 1970. Metapyrocatechase (*Pseudomonas*). In: Methods in Enzymology, **17a**, 522.

ORNSTON, L. N., 1966. The conversion of catechol and protocatechuate to β-ketoadipate by *Pseudomonas putida*. IV. Regulation. J. biol. Chemistry, **241**, 3800.

v. PLOTHO, O., 1950. Untersuchungen an Proactinomyceten. Arch. Mikrobiol., **14**, 12.

PRAUSER, H., 1981. Nocardioform organisms: general characterization and taxonomic relationships, pp. 18—24. In: K. P. SCHAAL, G. PULVERER (eds.), Actinomycetes. Zbl. Bakt. Suppl. 11. Gustav-Fischer Verlag, Stuttgart—New York.

RABSCH, W. und FRITSCHE, W., 1977. Physiologie des Anilinabbaus durch *Achromobacter* Ir 2. Z. Allg. Mikrobiol., **17**, 139.

RANN, D. L. and CAIN, R. B., 1969. The regulation of the enzymes of aromatic ring fission in an actinomycete. Biochem. J., **114**, 77 P.

REBER, H., HELM, V. and KARANTH, N. G. K., 1979. Comparative studies on the metabolism of aniline and chloroanilines by *Pseudomonas multivorans* strain An-1. Eur. J. appl. Microbiol. Biotechnol., **7**, 181.

REBER, H. and THIERBACH, G., 1980. Physiological studies on the oxidation of 3-chlorobenzoate by *Acinetobacter calcoaceticus*. Eur. J. appl. Microbiol. Biotechnol., **10**, 223.

SALA-TREPAT, J. M. and EVANS, W. C., 1971. The meta cleavage of catechol by *Azotobacter* species. 4-Oxalocrotonate pathway. Eur. J. Biochem., **20**, 400.

SUROVTSEVA, E. G. and VOLNOVA, A. I., 1980. Aniline as a sole source of carbon, nitrogen and energy for *Alcaligenes faecalis*. Mikrobiologiya, **49**, 49.

SUROVTSEVA, E. G. and VOLNOVA, A. I., 1981. 4-Chlorocatechol, an inhibitor of pyrocatechol 2,3-dioxygenase in *Alcaligenes faecalis*. Mikrobiologiya, **50**, 386.

SUROVTSEVA, E. G., VOLNOVA, A. I. and SHATSKAYA, T. Ya., 1980. Degradation of monochlorosubstituted anilines by *Alcaligenes faecalis*. Mikrobiologiya, **49**, 351.

THORNLEY, M. J., 1960. The differentiation of *Pseudomonas* from other gram-negative bacteria on the basis of arginine metabolism. J. appl. Bacteriol., **23**, 37.

WALKER, N. and HARRIS, D., 1969. Aniline utilization by a soil pseudomonad. J. appl. Bacteriol., **32**, 457.

WILLIAMS, P. A., CATTERALL, F. A. and MURRAY, K., 1975. Metabolism of naphthalene, 2-methylnaphthalene, salicylate, and benzoate by *Pseudomonas* PG: Regulation of tangential pathways. J. Bacteriol., **124**, 679.

YOU, I.-S. and BARTHA, R., 1982. Metabolism of 3,4-dichloroaniline by *Pseudomonas putida*. J. agric. Food Chem., **30**, 274.

Mailing address: Dr. D. JANKE
Zentralinstitut für Mikrobiologie
und experimentelle Therapie
der AdW
DDR-6900 Jena, Beutenbergstr. 11

Zeitschrift für Allgemeine Mikrobiologie	23	1983	4	247–257

(Sektion Biologie — Allgemeine Mikrobiologie — Friedrich-Schiller-Universität Jena)

Protoplastenfreisetzung und Regeneration des Ascomyceten *Hypomyces ochraceus* (PERS.) TUL.

K. H. RIEMAY, SILVIA ELLRICH und R. TRÖGER

(Eingegangen am 26. 7. 1982)

A rapid and convenient method for producing protoplasts from 3 d old mycelium of the ascomycete *Hypomyces ochraceus* is described. The procedure involves a *Helix pomatia* enzyme preparation and sucrose (20%) for stabilization. Pretreatment with disulfide bond reducing agents reduced the amount of viable protoplasts. Formation of protoplasts and different stages of regeneration were observed by phase contrast microscopy. There was only one type of true regeneration from protoplasts to hyphae in 15—30% gelatine-medium by direct forming a germ tube from the original protoplast. Cytological events and physiological conditions are discussed.

Pilzprotoplasten werden als geeignete Objekte zur Lösung unterschiedlicher Fragestellungen (Zellwandaufbau, Morphogenese, Membrantransport, Isolierung von Organellen, Fusion *etc.*) vielfach verwendet. Die Freisetzung von Protoplasten aus Hyphen-

Tabelle 1
Einwirkung von Schneckenenzym auf verschiedene Pilzarten zur Protoplastenfreisetzung Pilzstämme aus der Pilzkulturensammlung Weimar der FSU Jena)

Organismus	Protoplasten	Abweichungen von den unter MM beschriebenen Bedingungen bzw. cytologische Veränderungen
Sporodinia grandis p 73	+ (AP)	Proteinase (10 mg/ml)-Vorbehandlung des Mycels notwendig, Enzymkonzentration 100—150 mg/ml
Mucor spec. p 72	+	Protoplasten nur aus der Hyphenform, Hefeform auch nach Vorbehandlung (Thiolverbindungen, Enzyme) keine Protoplasten
Mycotypha microspora p 44	—	Nur Anschwellen der Hyphenspitzen, keine Protoplasten
Morchella elata m 441-9	+	Hyphen zerfallen, Protoplastenfreisetzung erst nach 24—48 h aus den angeschwollenen Enden der Hyphenbruchstücke
Verpa digitaliformis m 472-1	—	Auch nach Vorbehandlung, Erhöhen der Enzymkonzentration bis 150 mg/ml und Inkubation bis 36 h keine Protoplasten
Ptychoverpa bohemica m 465-8		
Hypomyces ochraceus m 359	+	mehr als 10^7 Protoplasten/ml
Hypomyces odoratus m 329	+	*ca.* 7 × 10^6 Protoplasten/ml
Hypomyces rosellus m 58	+	*ca.* 10^5 Protoplasten/ml
Piptoporus betulinus p 100-5	+ (AP)	Freisetzung aus Spitzen und lateralen Hyphenabschnitten gleichzeitig, Beginn nach 10 min, bis 120 min Inkubation nur wenige, kleine (3—6 μm) Protoplasten
Coprinus comatus c 20-2	—	bis 72 h Inkubation keine Protoplasten
Coprinus cinereus c 26-1	—	
Fusarium decemcellulare i 57 Konidien	—	bis 200 mg Enzym/ml und 48 h Inkubation mit Vorbehandlung nur Austritt von Vakuolen an Konidienenden

AP — Ergebnisse nur an Agarfilmpräparaten reproduziert

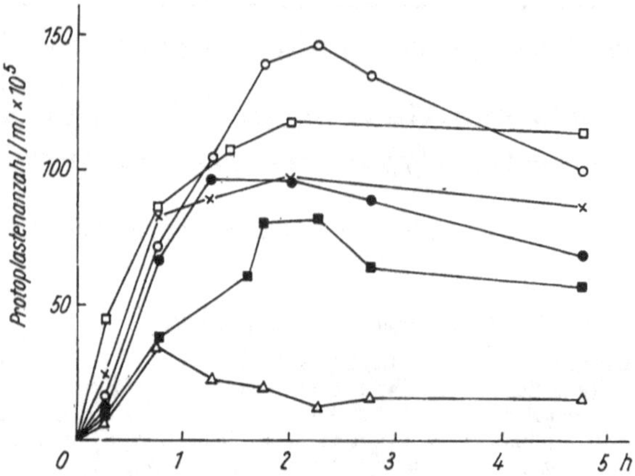

Abb. 1. Protoplastenausbeute bei Verwendung verschiedener Stabilisatoren (Mycelalter 3 d): □ NaCl 0,8 M, △ KCl 0,8 M, ■ $MgSO_4$ 1,0 M, ○ Saccharose 20%, × Mannit 0,8 M, ● Sorbit 1,0 M (h — Einwirkungszeit des Enzyms in Stunden)

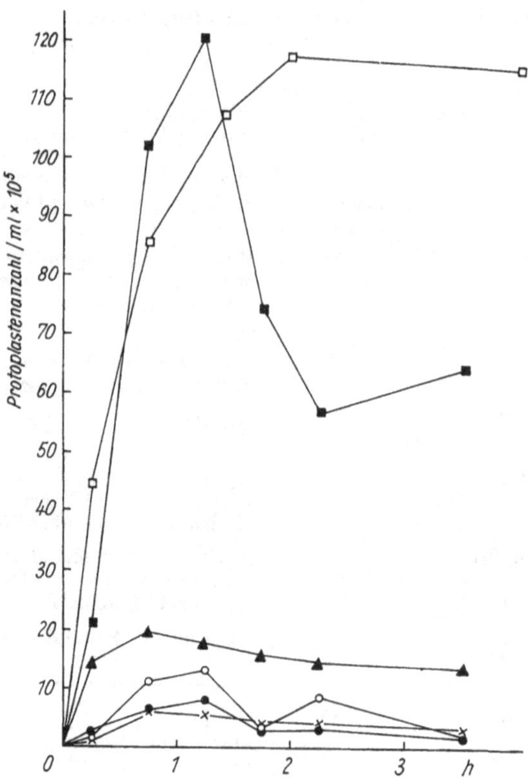

Abb. 2. Protoplastenausbeute von Mycel (Alter 3 d) aus verschiedenen Kulturmedien: □ NEČAS, × HENNEBERG mit 50 g Glucose/l, ■ HENNEBERG mit 30 g Glucose/l, ○ RAULIN-THOM, ● RAULIN-THOM mit Zusatz von 5 g Pepton/l (h — wie in Abb. 1)

pilzen durch zellwandlytische Enzyme, Inhibierung der Zellwandbildung und autolytische Enzympräparate beschrieben zusammenfassend STRUNK (1970), VILLANUEVA und GARCIA-ACHA (1971) und PEBERDY (1972, 1978, 1979). Trotz umfangreicher Kenntnisse über Bildung, Eigenschaften und Verwendung von Protoplasten ist für ein gegebenes neues Objekt häufig eine spezielle Adaptation bekannter Methoden notwendig (Unterschiede im Zellwandbau) und nicht immer sofort erfolgreich. In der vorliegenden Arbeit werden Ergebnisse zur Freisetzung und Regeneration von Protoplasten aus *Hypomyces ochraceus* m 359 vorgelegt. Dieser Pilz erwies sich nach Untersuchung von 13 Hyphenpilzen (3 Zygomyceten, 7 Ascomyceten, 3 Basidiomyceten, s. Tab. 1) als besonders geeignet zur Herstellung stabiler Protoplasten für physiologische Arbeiten.

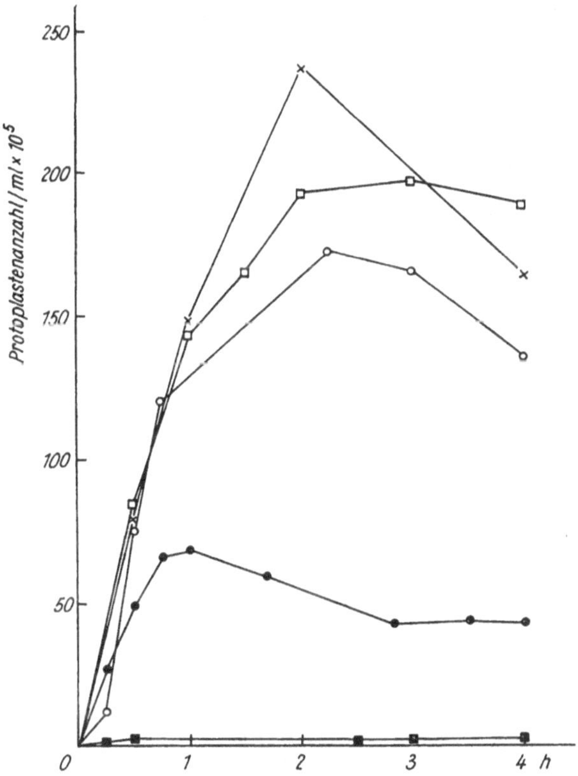

Abb. 3. Protoplastenausbeute nach 3tägiger Kultivierung des Mycels mit verschiedenen C-Quellen: Raulin-Thom-Medium mit □ Saccharose, ■ Glucose, ▲ Melibiose (je 75 g/l); Henneberg-Medium mit ○ Saccharose, ● Glucose, × Melibiose (je 30 g/l) (h — wie in Abb. 1)

Material und Methoden

Organismus und Medien: Die Anzucht des Ascomyceten *Hypomyces ochraceus* (Pers.) Tul. Stamm m 359 erfolgte in Kulturröhrchen mit Schrägagar bzw. in Petrischalen (Durchmesser 100 mm) auf festem Nährmedium: 30 g Malzextrakt, 2 g Pepton, 30 g Agar, 1000 ml Leitungswasser pH 6,8. Die Sterilisationszeit im Autoklaven betrug 30 min bei 121 °C. Die Kultivierung zur Mycelgewinnung fand in flüssigem Medium nach Henneberg (Janke, 1946; verändert) statt: 5 g Pepton, 5 g $NH_4H_2PO_4$, 2 g KNO_3, 0,5 g $MgSO_4 \cdot 7\,H_2O$, 0,1 g $CaCl_2$, 30 g Glucose, 1000 ml aqua dest. pH 5,0. Für die mikroskopische Beobachtung kamen die Medien nach Nečas (1956, 1962) zur Anwendung: 50 g Saccharose, 20 g Asparagin, 25 g KH_2PO_4, 5 g $MgSO_4 \cdot 7\,H_2O$, 1000 ml Leitungswasser, bzw. 5 g Saccharose, 1,25 g Asparagin, 1,25 g KH_2PO_4, 0,425 g $MgSO_4$, 50 ml aqua dest., 2% Agar (pH 5,4). Sterilisation fraktionisiert an 3 aufeinanderfolgenden Tagen jeweils 20 min bei 100 °C. Für die Regenerationsversuche mit Pilzprotoplasten in flüssigem Medium verwendeten wir die Nährlösung nach Raulin-Thom (Janke 1946). Zur Herstellung eines Nährbodens wurde diese mit 15 bzw. 30% Gelatine oder mit 3% Agar verfestigt.

Enzympräparation: Das Schneckenenzym präparierten wir aus dem Eingeweidesack von *Helix pomatia*, filtrierten und trockneten lyophil (Aufbewahrung bei —4 °C).

Protoplastenherstellung: Zur Herstellung der Protoplasten modifizierten wir eine Methode von STRUNK (1965, 1969). Dabei wurde ein 500-ml-Stehkolben mit 100 ml RAULIN-THOM-Lösung mittels Mycelstück (Malzagarbewuchs, Kolonierandzone) beimpft und geschüttelt. Nach 3 Tagen erfolgte die Mycelernte. Anschließend wurden die Hyphen mit Stabilisatorlösung (20% Saccharose in 0,04 M Phosphatpuffer, pH 7,0) gewaschen. Im Inkubationsansatz (25-ml-Erlenmeyerkolben) wurden 80 mg frisches Mycel/ml mit 20—25 mg/ml Schneckenenzym versetzt, bei 23 °C geschüttelt (Schüttelfrequenz 150/min) und die Protoplasten nach 90—120 min (oder kürzeren Zeiten) durch Wattefiltration abgetrennt. Für physiologische Arbeiten wurden die Protoplasten durch Gradientenzentrifugation (1 M $MgSO_4$-Lösung, überschichtet mit 0,7 M NaCl, 3000 × g 10 min) gereinigt und konzentriert. Die Reinheit wurde stets im Lysetest (s. Regeneration) geprüft. Das Waschen

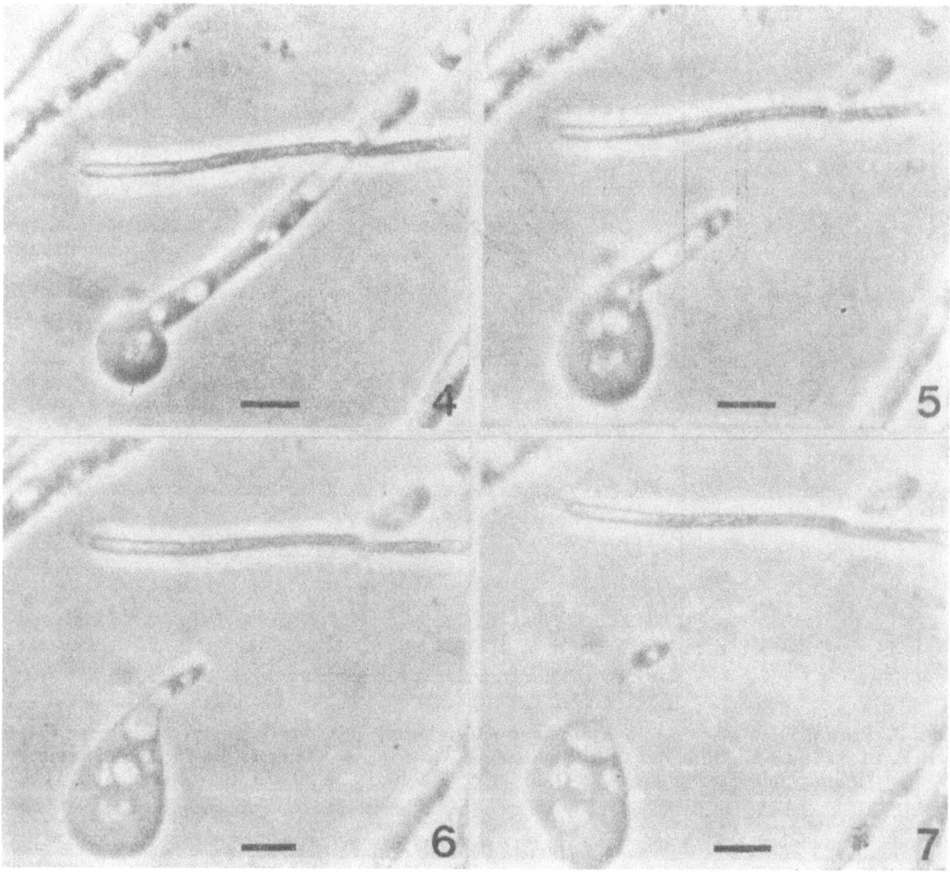

Abb. 4.—7. Protoplastenbildung im Agarfilmpräparat, Aufnahmen im Abstand von 1 min; in Abb. 5—7 leeres Hyphenkompartiment mit vollständig erhaltener Hyphenwand nach Ausschlüpfen des Protoplasten sichtbar. — 10 μm

und Aufbewahren der Protoplasten erfolgte in Stabilisatorlösung: 0,7 M NaCl oder 20% Saccharose in 0,04 M Phosphatpuffer, pH 7,0. Als günstigste Aufbewahrungstemperatur erwies sich 4 °C. Die Protoplasten-Anzahl wurde unter Verwendung einer THOMA-Zählkammer ermittelt.

Regeneration: Für die Regenerationsversuche kam 0,1 ml der hergestellten Protoplastensuspension zur Anwendung, die in Agar- oder Gelatinefilm (auf Objektträgern) eingebracht wurde (feuchte Kammer, 23 °C). Die Untersuchungen zur Regeneration in flüssigen Medien erfolgten auch mit NEČAS-Medium. Bei allen Regenerationsversuchen dienten 0,1 ml der Protoplastensuspension in 2 ml aqua dest. als Kontrolle. Es trat sofortige Lyse der Protoplasten ein. Von dieser Lösung brachten wir Tropfen auf bzw. in Nährmedien, so daß eine eventuelle Infektion der Protoplastensuspension durch Pilzsporen oder Hefen nach der Inkubation sichtbar wurde und eine Verwechslung mit regenerierenden Protoplasten ausgeschlossen war. Die mikrofotografischen Aufnahmen erfolgten mittels Hellfeld- bzw. Phasenkontrastverfahren und ORWO NP 15 unter Verwendung von Agar- oder Gelatinefilm-Präparaten.

Ergebnisse und Diskussion

Protoplastenbildung

Die Abbildungen 4—7 und 8—11 zeigen typische Abläufe der Protoplastenbildung, die stets an Hyphenspitzen erfolgte und bereits wenige Minuten nach Inkubation der Hyphen mit Schneckenenzym begann. Eine Auflösung der Zellwand auf der gesamten Hyphenoberfläche konnten wir im Gegensatz zu YAKOVLEV u. YAVAKHIYA (1981) und

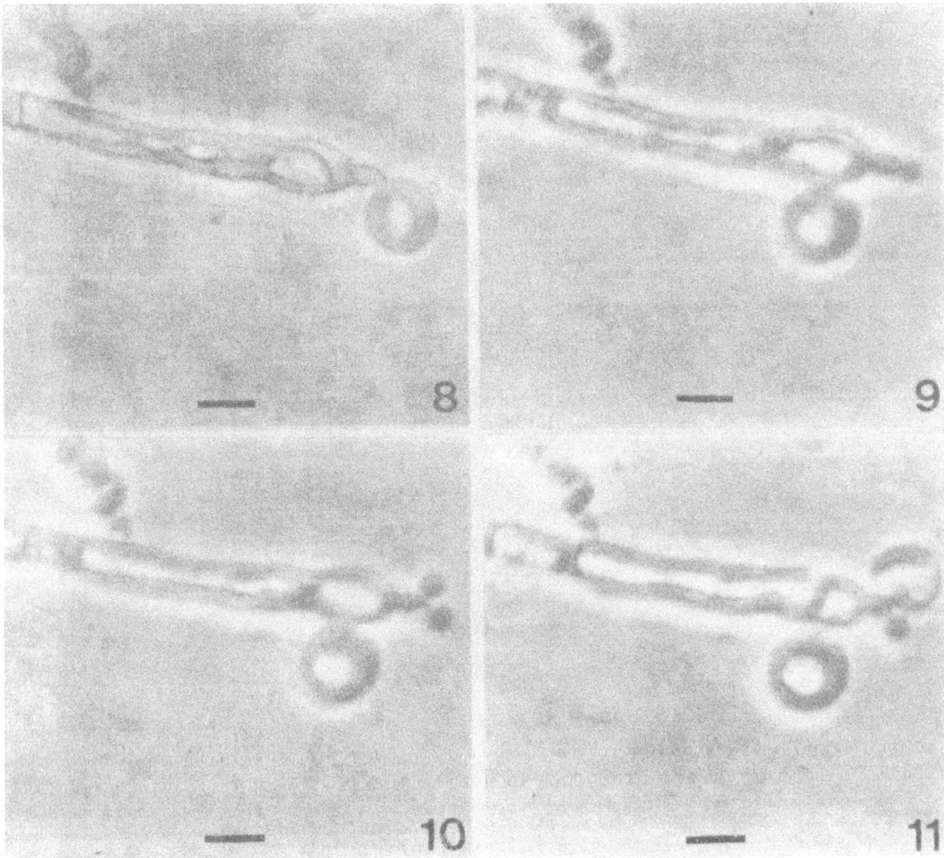

Abb. 8.—11. Entstehung mehrerer Protoplasten unterschiedlicher Größe aus einem Hyphenkompartiment, Agarfilmpräparat, Aufnahmen im Abstand von 2 min. — 10 µm

TANAKA *et al.* (1981) nur vereinzelt in Agarfilmpräparaten beobachten. Die Hyphen bildeten dabei keine Protoplasten. Wir führen dies auf eine lokal zu hohe Enzymkonzentration zurück, die Membranen und Protoplasma schädigt. Die bei normalem Verlauf austretenden Protoplasten waren inhomogen sowohl hinsichtlich ihrer Größe (nicht vakuolisierte Protoplasten 3—15 µm, vakuolisierte 10—40 µm Durchmesser), als auch der Kernzahl (0 bis 5 Kerne je Protoplast in elektronenmikroskopischen Auf-

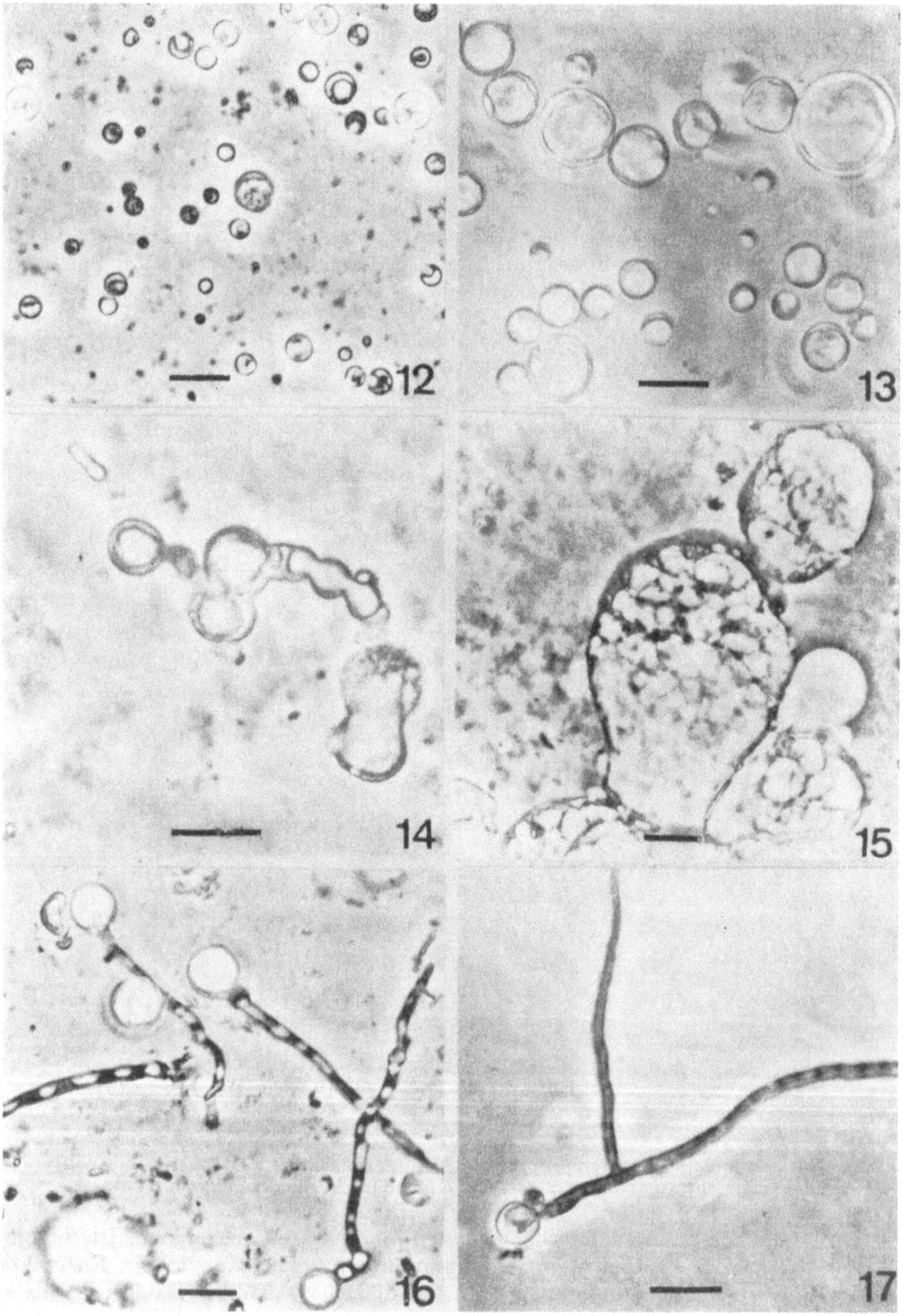

Abb. 12 – Abb. 17

nahmen). Protoplastenaustritt aus lateralen Hyphenabschnitten trat erst nach 60 bis 90 min Inkubation und seltener als an Hyphenspitzen auf. Die Abbildungen 5—7 zeigen die vollständig erhaltene Hyphenwand, die nach dem Schlüpfen der Protoplasten zurückbleibt. Der Hyphendurchmesser betrug hier *ca.* 5,5 µm, der Protoplast maß *ca.* 19 × 27 µm (Agarfilmpräparat, verzögerte Abrundung).

Enzymwirkung

Bei Einsatz von 80 mg Feuchtmycel und 20—25 mg Schneckenenzym pro ml Inkubationsansatz, unter Verwendung von 0,7 M NaCl bzw. 20% Saccharose (*ca.* 0,6 M) als Stabilisator, entstanden nach 90—105 min mehr als 10^7 Protoplasten pro ml. Diese Ausbeute variierte bei Verwendung unterschiedlicher Schneckenenzym-Präparate von 3×10^6 bis 1×10^7 Protoplasten pro ml Inkubationsansatz. Das Enzymgemisch unseres eigenen Präparates war über einen cH^+-Bereich von pH 4 bis pH 7 wirksam. Inkubationstemperaturen von 18 bis 35 °C ergaben stabile Protoplasten. Bei Temperaturen über 40 °C hergestellte Protoplasten lysierten um so rascher, je höher die Inkubationstemperatur lag.

Die Anwendung des von Popov *et al.* (1980) zur Herstellung von *Candida*-Protoplasten benutzten *Arthrobacter*-Enzympräparates, eines Brauereienzym-Gemisches (1000 IE β-Glucanase, 150 TE Protease, 3000 IE Amylase), von Proteinase K (Serva) bzw. Trypsin (Fa. Leitholdt, DDR) allein, oder in verschieden dosierten Kombinationen, ergab unter unseren Bedingungen keine Protoplasten. Verantwortlich dafür dürfte u. a. die geringe (*Arthrobacter*-Enzym) bzw. fehlende Chitinase-Aktivität sein. *H. ochraceus* weist, verglichen mit anderen Hyphenpilzen, einen hohen Chitinanteil in der Zellwand auf (Riemay, unveröff.). Trypsin führte bei Inkubationszeiten von 48—72 Stunden zur Schädigung der gesamten Zelle (granuliertes Protoplasma). Vorbehandlung des Mycels mit 2-Mercaptoäthanol (10, 50, 100 bzw. 500 mM) mit und ohne EDTA-Zusatz, oder mit Dithioerythrit (Serva) erniedrigte die Protoplastenausbeute. Fawcett *et al.* (1973) und Anne *et al.* (1974) hatten bei *Penicillium* ebenfalls keine Begünstigung der Protoplastenbildung durch Thiolverbindungen gefunden, während *Cephalosporium* nur bei Vorbehandlung Protoplasten bildete. Finkelman *et al.* (1980) behandelten *Aureobasidium*, Yakovlev u. Yavakhiya (1981) *Pyricularia* mit 2-Mercaptoäthanol und verkürzten bei gleichzeitiger Erhöhung der Ausbeute die Inkubationszeit. Diese aus dem unterschiedlichen Zellwandaufbau der Hyphenpilze erklärbaren Differenzen sind eine Ursache für die in der Literatur beschriebenen Variationen der Protoplastenmethodik und die Notwendigkeit ihrer Anpassung bei noch nicht bearbeiteten Arten.

Einfluß des Stabilisators, des Anzuchtmediums und des Mycelalters

Abbildung 1 zeigt den Einfluß organischer und anorganischer Stabilisatoren auf die Protoplastenausbeute. Saccharose wurde in Endkonzentrationen von 10% (*ca.* 0,3 M,

Abb. 12. Protoplastensuspension nach dem Schlüpfen

Abb. 13. Lebende Protoplasten (regenerationsfähig) nach 14tägiger Haltung in gepufferter Saccharoselösung (20%, pH 7,0) bei 4 °C, stark vakuolisiert und vergrößert

Abb. 14. Irreguläre Regeneration ohne Hyphenbildung in Raulin-Thom-Medium mit 3% Agar, nach 18 h

Abb. 15. Ausbildung von Riesen-Protoplasten nach 18 h Regeneration in Malz-Pepton-Agar, keine Hyphenbildung

Abb. 16. — 17. Regenerierende Protoplasten in Raulin-Thom-Medium mit 30% Gelatine nach 20 h, Hyphenbildung und -verzweigung

Abb. 12—17: ——— 20 µm

Ausbeute 2×10^5 Protoplasten/ml), 15% (*ca.* 0,44 M, Ausbeute 2×10^6/ml) und 20% (*ca.* 0,6 M, Ausbeute $6{,}5 \times 10^6$ bis $1{,}9 \times 10^7$/ml) verwendet. Eine Vorbehandlung des Mycels (2 h, 4 h) mit Stabilisatoren (NaCl, Saccharose) vor der Inkubation mit dem Enzym erniedrigte die Protoplastenausbeute.

Die Aussagen zur Brauchbarkeit von Stabilisatoren differieren in der Literatur mit den verwendeten Pilzen sehr stark. Während sich für *Hypomyces* NaCl (0,7–0,8 M) und Saccharose (0,6 M) als besonders geeignet, Mannit, Sorbit und $MgSO_4$ weniger und KCl als ungeeignet erwiesen, verwendeten Fawcett *et al.* (1973) nur anorganische Salze (NaCl, KCl, NH_4Cl, 0,6–0,9 M) mit Erfolg. Sorbit, Mannit bzw. Saccharose ergaben bei *Penicillium* sehr wenige, bei *Cephalosporium* bedeutend weniger Protoplasten als die anorganischen Stabilisatoren. Troitskij *et al.* (1978) verwendeten bei *Sclerotinia* Saccharose (20%), Yakovlev u. Yavakhiya (1981) dagegen bei *Pyricularia* KCl bzw. $(NH_4)_2SO_4$ (beide 0,4 M) und Stahl *et al.* (1977) bei *Claviceps* NaCl und KCl (0,9 M) mit Erfolg zur Protoplastenstabilisierung. Die Verallgemeinerung von Peberdy (1979), alle anorganischen Salze seien bei Hyphenpilzen effektivere Stabilisatoren der Protoplastenbildung, trifft für unser Objekt nicht zu. NaCl und KCl hatten entgegengesetzte Wirkung auf die Stabilität der entstehenden Protoplasten, Saccharose war noch besser geeignet als NaCl.

Problematisch war die geringe Mycelausbeute mit Raulin-Thom-Medium (nur 115 mg Frischmycel pro Kolben nach 3 d). Da Henneberg-Medium einen 10fach höheren Mycelertrag, aber niedrige Protoplastenerträge ergab, suchten wir durch Veränderungen des Mediums die Erträge zu erhöhen. Die Abbildungen 2 und 3 zeigen, wie die Protoplastenbildung durch das Anzuchtmedium, und nach unseren Beobachtungen besonders durch Art und Konzentration der C-Quelle, beeinflußt wird. So konnte bereits mit einer Erhöhung der Glucosekonzentration von 30 auf 50 g/l im Anzuchtmedium die Protoplastenbildung signifikant gesteigert werden (Abb. 2). Möglicherweise wird der Glucananteil der Zellwand unter diesen Bedingungen gegenüber Chitin so erhöht, daß die besonders aktiven Glucanasen des Schneckenenzyms die Wand schneller abbauen können. Eine Veränderung der N-Quelle im Raulin-Thom-Medium durch Zugabe von Pepton (5 g/l, von *Hypomyces* sonst gut verwertet) erniedrigte dagegen die Protoplastenbildung (Erhöhung des Chitinanteils der Wand).

Das Alter des verwendeten Mycels beeinflußte die Protoplastenfreisetzung nachhaltig. Unter gleichen Bedingungen erhielten wir nach 2 h Inkubation mit Schneckenenzym aus 3 d altem Mycel (lineare Wachstumsphase, erstes Drittel) maximal $1{,}9 \times 10^7$ Protoplasten pro ml, aus 5 d Mycel (lineare Wachstumsphase, zweites Drittel) 6×10^6/ml und aus 17 d Mycel (Autolysephase) nur noch 5×10^5 Protoplasten pro ml. Eine Förderung der Protoplastenbildung durch Beteiligung endogener lytischer (autolytischer) Enzyme, wie sie Peberdy (1979) aus den Versuchen von Anne *et al.* (1974) folgerte, liegt offenbar nicht vor, obwohl Enzympräparate aus autolysierenden Kulturen von *Hypomyces ochraceus* aus 3 d altem Mycel des gleichen Pilzes haltbare Protoplasten freisetzen können (Riemay, unveröff.). Die Beobachtung von Polacheck u. Rosenberger (1975), daß ältere Hyphenwände durch Hydrolasen schwerer abbaubar sind als jüngere, läßt die mit zunehmendem Mycelalter abnehmende Protoplastenausbeute verständlich werden.

Haltbarkeit der Protoplasten

Eine frisch hergestellte Protoplastensuspension (Abb. 12) enthält Protoplasten verschiedener Größe, Kernzahl (s. Protoplastenbildung), Vakuolisierung und Lichtbrechung. Nicht vakuolisierte Protoplasten waren kleiner (bis 0,5 µm) und erschienen im Phasenkontrast dunkler, als die größeren Protoplasten mit einer bis mehreren Vakuolen. Nach längerer Aufbewahrung im gepufferten Saccharosemedium (pH 7) bei 4 °C schwollen die Protoplasten durch Flüssigkeitsaufnahme so stark an, daß ihr Protoplasma nur noch als dünner Film auf der Innenseite der Membran auflag (Abb. 13). Nach 4 bis 6 Wochen war die Protoplastenzahl zwar deutlich zurückgegangen (60—80% Überlebende nach 4 Wochen), die Lebensfähigkeit der Überlebenden aber nicht beeinträchtigt. In gepufferter NaCl (0,7 M) waren schon nach 7—10 Tagen alle Protoplasten geplatzt, obwohl NaCl bei der Herstellung der Protoplasten der Saccharose gleichwertig war. Saccharose ist u. U. für eine längere Haltung der Protoplasten geeignet, weil sie von den Protoplasten zwar sehr langsam, aber doch als Energiequelle genutzt werden kann. Die Ergebnisse zeigen, daß Stabilisator, Anzuchtbedingungen, Mycelalter und Inkubationsbedingungen nicht nur die Bildung, sondern auch Lebensdauer und Vitalität der Protoplasten und damit deren Regenerationsfähigkeit entscheidend mitbestimmen.

Regeneration

Die Untersuchungen zur Regeneration der Protoplasten ergaben eine deutliche Abhängigkeit des Regenerationsverlaufs vom verwendeten Medium. Mit RAULIN-THOM-Medium, verfestigt durch 2 bzw. 3% Agar, war es nicht möglich, aus den Protoplasten normales Mycel zu erhalten. Nach 18 h hatten sich die Protoplasten vergrößert und durch einen der Sprossung ähnlichen Vorgang vermehrt (Abb. 14), wobei auch längere Zellketten entstanden. Die Vermehrung erfolgte durch eine Membranausstülpung, die sich vergrößerte und in die dann der Inhalt des Protoplasten übertrat. Auch nach 72—96 h konnte keine Hyphenbildung beobachtet werden. TANAKA *et al.* (1981) bezeichneten eine ähnliche, von ihnen in flüssigem Regenerationsmedium beobachtete Kettenbildung als „incomplete regeneration" (Typ 1). Im Gegensatz zu ihnen konnten wir jedoch nie ein Auswachsen von Hyphen aus solchen Zellketten beobachten, das die Autoren als Regeneration vom Typ 2 bezeichneten. Versuche, Protoplasten auf Agaroberflächen zu regenerieren, blieben erfolglos. Die Protoplasten autolysierten stets nach vorheriger sehr starker Schwellung. Eine Vermehrung trat selten auf. Ersetzten wir RAULIN-THOM-Agar durch Malzextrakt-Pepton-Agar, so bildeten die Protoplasten Riesenformen. Nach 18 h erreichten sie infolge Vermehrung und Vergrößerung der Vakuolen die 2- bis 3fache Größe (Abb. 15). Nach 42 h bedeckten riesige, unregelmäßig geformte Protoplasten mit einem Durchmesser bis über 150 µm die gesamte Agaroberfläche des Präparates. Sprossung oder Hyphenbildung trat nicht auf. Auch YAKOVLEV u. YAVAKHIYA (1981) erreichten mit Agargelen keine Regeneration, während TROITSKIJ *et al.* (1978) erfolgreich CZAPEK-DOX-Agar mit 20% Saccharose verwendeten. *Hypomyces*-Protoplasten regenerierten nur in Gelatinegelen bis zur Hyphenform. In RAULIN-THOM-Medium mit 15% Gelatine regenerierten die Protoplasten unter *de novo* Synthese einer Hyphenwand.

Die meist stark vakuolisierten Hyphen wuchsen ohne Sprossungsvorgänge direkt aus dem Protoplasten heraus. Auch in Gelatinegelen traten (1) Vermehrung der Protoplasten durch Sprossung, (2) Kettenbildung und (3) unregelmäßige Wuchsformen mit starker Schwellung durch zunehmende Vakuolisierung ohne Sprossung auf. Diese Vorgänge (1—3) führten aber nie zur Ausbildung von Hyphen mit Spitzenwachstum, wie es ANNE *et al.* (1974), RAMOS u. GARCIA-ACHA (1975), TROITSKIJ *et al.* (1978) und TANAKA *et al.* (1981) von sprossenden oder kettenbildenden Protoplasten beschrieben. Nach VALK u. WESSELS (1976, 1977) kann angenommen werden, daß bei allen unvollständigen Regenerationsformen die Zellwand-Synthese nur teilweise oder gestört abläuft. Sproßzellen (aus Zellketten) bzw. Hyphen weisen unterschiedliche Muster der Baustein-Anlagerung und des Aufbaus in der Wand auf. Durch Erhöhen der Gelatinekonzentration auf 30% konnte die Regenerationsrate erheblich gesteigert werden. Etwa 80—90% der Protoplasten regenerierten bis zur Hyphenform, wobei in sehr unterschiedlichem Grade vakuolisierte Protoplasten gleichartig regenerierten. Die Hyphen (Abb. 16—18) erreichten hier, verglichen mit der Regeneration in 15%igen Gelatinegelen nach 20 bzw. 66 h die 2- bis 3fache Länge. Vermehrung der Protoplasten durch Sprossung trat nicht mehr auf. Hyphen entstanden immer direkt aus den Protoplasten, ohne daß diese eine sichtbare Wandverdickung aufwiesen, wie sie Peberdy (1978) und Tanaka *et al.* (1981) beschrieben. Verzweigungen (Abb. 17) und starke Vakuolisierung (Abb. 18) traten an sehr jungem Mycel auf. An der Oberfläche von Gelatinegelen blieben die Protoplasten zwar einige Tage stabil und vergrößerten sich, zeigten aber keine Vermehrung oder Hyphenbildung. Schocklyse durch Überschichten mit aqua dest. ließ keine Zellwandrückstände erkennen. Erhöhte Saccharosekonzentrationen (über 7%) im Regenerationsmedium mit 30% Gelatine hemmten die Regeneration; nur noch 10% der Protoplasten regenerierten bis zur Hyphenform.

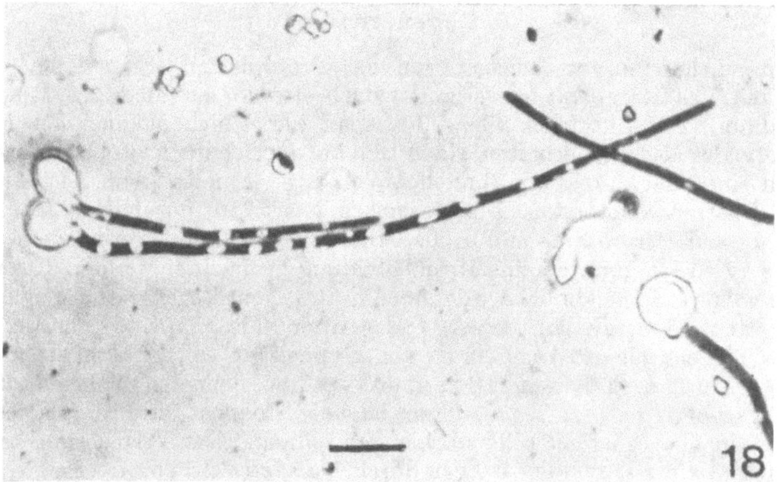

Abb. 18. Regenerationsstadien in Raulin-Thom-Medium mit 30% Gelatine nach 20 h, Hyphenbildung direkt aus dem Protoplasten, Doppelkonturen der Protoplasten infolge innen aufliegenden dünnen Protoplasmafilms, — 20 μm

Besondere Bedeutung für die Vollständigkeit des Regenerationsablaufes hatte die Sauerstoffversorgung. Wurden Präparate unter sonst optimalen Bedingungen zum Schutz vor Austrocknung mit Vaseline umrandet, so sank die Regenerationsrate rasch ab, während nicht umrandete Kontrollen in sterilen feuchten Kammern zu 80 bis 90% regenerierten. Regenerationsversuche unter Verwendung flüssiger Medien führten zu ähnlichen Stadien, wie bei Kultivierung mit 15% Gelatine; eine Hyphenbildung trat jedoch nicht auf. Bei der Bildung von Zellketten blieben nach dem Übertritt des Protoplasmas in die jeweils neu gebildete Ausstülpung (Sproßzelle) leere Membranen zurück, die in aqua dest. lysierten. Vereinzelt regenerierten Protoplasten im Aufbewahrungsmedium (gepufferte Saccharoselösung) bei 4 bis 8 °C nach vorhergehender Aggregation zu Clustern und der Ausbildung einer dicken gelartigen äußeren Matrix (vergleichbar der von Tanaka *et al.* (1981) beschriebenen, auf der Protoplastenoberfläche gebildeten aberranten Zellwand; Regenerationstyp 3), zu sehr dünnen (2 bis 3,5 μm) Hyphen. Alle Ergebnisse wurden an frischen Protoplastensuspensionen erhalten; eine 4- bis 6tägige Aufbewahrung bei 4 °C hatte keine negativen Auswirkungen auf den Regenerationsverlauf. Nach 10tägiger Aufbewahrung regenerierten aber nur noch etwa 10% der Protoplasten zu Hyphen, wobei die ersten Hyphen erst nach 30 bis 40 h auftraten.

Literatur

ANNE, J., EYSSEN, H. and DE SOMER, P., 1974. Formation and regeneration of *Penicillium chrysogenum* protoplasts. Arch. Mikrobiol., **98**, 159—166.

FAWCETT, P. A., LODER, P. B., DUNCAN, M. J., BEESLEY, T. J. and ABRAHAM, E. P., 1973. Formation and properties of protoplasts from antibiotic-producing strains of *Penicillium chrysogenum* and *Cephalosporium acremonium*. J. gen. Microbiol., **79**, 293—309.

FINKELMAN, M. A. J., ZAJIC, J. E. and VARDANIS, A., 1980. New method of producing protoplasts of *Aureobasidium pullulans*. Appl. Environ. Microbiol., **39**, 923—925.

JANKE, A., 1946. Arbeitsmethoden der Mikrobiologie, Bd. 1. Th. Steinkopff Dresden-Leipzig.

NEČAS, O., 1956. Regeneration of yeast cells from naked protoplasts. Nature (Lond.), **177**, 898 to 899.

NEČAS, O., 1962. The mechanism of regeneration of yeast protoplasts. II. Physical conditions. Folia biol., **8**, 256—262.

PEBERDY, J. F., 1972. Protoplasts from fungi. Sci. Progress Oxford, **60**, 73—86.

PEBERDY, J. F., 1978. Protoplasts and their development. In: "The Filamentous Fungi", Vol. 3 (ed. by J. E. SMITH and D. R. BERRY). London Edward Arnold, pp. 119—131.

PEBERDY, J. F., 1979. Fungal protoplasts: Isolation, reversion and fusion. Ann. Rev. Microbiol., **33**, 21—39.

POLACHECK, Y. and ROSENBERGER, R. F., 1975. Autolytic enzymes in hyphae of *Aspergillus nidulans*: Their action on old and newly formed walls. J. Bacteriol., **121**, 332—337.

POPOV, B., REUTER, G. and MEYER, H. W., 1980. Cell wall regeneration of *Candida spec.* protoplasts. Z. Allg. Mikrobiol., **20**, 47—62.

RAMOS, S. and GARCIA-ACHA, I., 1975. Cell wall enzymatic lysis of the yeast form of *Pullularia pullulans* and wall regeneration by protoplasts. Arch. Microbiol., **104**, 271—277.

STAHL, CH., NEUMANN, D., SCHMAUDER, H. P. und GRÖGER, D., 1977. Protoplastengewinnung bei Mutterkornpilzen. Biochem. Physiol. Pflanzen, **171**, 363—368.

STRUNK, CH., 1965. Protoplasten-Entstehung und -Reversion bei Basidiomyceten (*Polystictus versicolor*). In: „Symposium über Hefe-Protoplasten", Jena 1965 (Hrsg. R. MÜLLER). Berlin Akademie Verlag 1977, S. 213—215.

STRUNK, CH., 1969. Licht- und elektronenmikroskopische Untersuchungen an regenerierenden Protoplasten von *Polystictus versicolor*. Z. Allg. Mikrobiol., **9**, 205—216.

STRUNK, CH., 1970. Protoplasts of moulds — the present state of knowledge. In: "Proc. 2nd Int. Symp. Yeast Protoplasts, Brno 1968", Acta Fac. Med. Univ. Brno, **37**, 107—119.

TANAKA, H., OGASAWARA, N. and KAMIMIYA, S., 1981. Protoplasts of *Pyricularia oryzae* P_2: Preparation and regeneration into hyphal form. Agric. Biol. Chem., **45**, 1541—1552.

TROITSKIJ, N. A., YAKOVENKO, K. N. and KOZLOVA, L. S., 1978. Production and regeneration of protoplasts of *Sclerotinia sclerotiorum* (LIB). DE BARY. Mikol. Fitopatol. (Leningrad), **12**, 475 to 479.

VALK, P. VAN DER and WESSELS, J. G. H., 1976. Ultrastructure and localization of wall polymers during regeneration and reversion of protoplasts of *Schizophyllum commune*. Protoplasma, **90**, 65—87.

VALK, P. VAN DER and WESSELS, J. G. H., 1977. Light and electronmicroscopic autoradiography of cell-wall regeneration by *Schizophyllum commune* protoplasts. Acta Bot. Neerl., **26**, 43—52.

VILLANUEVA, J. R. and GARCIA ACHA, I., 1971. Production and use of fungal protoplasts. In: "Methods in microbiology", Vol. 4 (ed. by C. BOOTH). Academic Press New York Lond., pp. 666—718.

YAKOVLEV, A. G. and YAVAKHIYA, V. G., 1981. Isolation and regeneration of mycel protoplasts of *Pyricularia oryzae* CAV. Mikol. Fitopatol. (Leningrad), **15**, 472—477.

Anschrift: Dr. K. H. RIEMAY
Friedrich-Schiller-Universität
Jena
Sektion Biologie
WB Allgemeine Mikrobiologie
DDR-6900 Jena, Neugasse 24

Zeitschrift für Allgemeine Mikrobiologie	23	1983	4	259–268

(Institut für technische Chemie der Akademie der Wissenschaften der DDR, Leipzig,
Direktor: Prof. Dr. sc. nat. M. RINGPFEIL)

Effektorwirkung von sauerstoffhaltigen C_1-Verbindungen bei der Kultivierung des methanotrophen Bakterienstammes GB 25

J. D. SCHNEIDER, K.-D. WENDLANDT, E. BRÜHL und G. MIRSCHEL

(Eingegangen am 30. 7. 1982)

Addition of oxygen-containing C_1-compounds to chemostat cultures of GB 25 increases both the yield of biomass and the specific growth rate. At optimum concentrations the catalytic activity of these compounds increases with increasing growth rates. Their influence on maintenance coefficients and maximum yield coefficients decreases in the order $CH_3OH > CO_2 > HCOOH > HCHO$. This result together with spectrophotometric NADH determinations suggests that the NADH pool determines the balance between the assimilatory and oxidative utilization of formaldehyde.

Methanotrophe Bakterien oxidieren Methan enzymatisch stufenweise über C_1-Intermediate zu Kohlendioxid (BROWN *et al.* 1964, JOHNSON u. QUAYLE 1964, BABEL u. MOTHES 1978) (Schema 1):

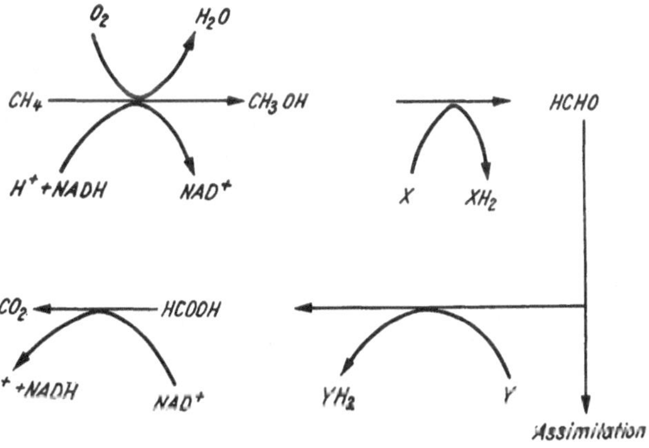

Bei Bakterien mit Serin-Zyklus erfolgt die Assimilation des Formaldehyds via Transformylierung an Glycin. Der von uns eingesetzte Bakterienstamm GB 25 assimiliert Formaldehyd bzw. aktivierten Formaldehyd via Serinweg (BABEL u. MIETHE, pers. Mitteilung). Es konnte sowohl bei methan- als auch bei methanolutilisierenden Mikroorganismen gezeigt werden, daß C_1-Intermediate das Wachstum beeinflussen können (PILAT u. PROKOP 1975, RICHTER *et al.* 1974, WENDLANDT u. BRÜHL 1980, PAPDUTSAKIS *et al.* 1981). Auch bei dem methanassimilierenden Bakterienstamm GB 25 bewirken die sauerstoffhaltigen C_1-Verbindungen Methanol, Formaldehyd, Ameisensäure und Kohlendioxid einen katalytischen Effekt, welcher zur Erhöhung der Zellausbeute und der maximalen Wachstumsrate führt (SCHNEIDER *et al.* 1983).

Eine optimale stimulierende Wirkung der C_1-Verbindungen wird bei einer Zugabe von *ca.* 2×10^{-2} g C_1-Verbindung/g gebildete Biomasse (als BTS) erreicht. Die katalytische Wirkung ist von der chemischen Natur der zugesetzten C_1-Verbindung abhängig, für den Stamm GB 25 in der Reihenfolge: $CH_3OH > CO_2 > HCOOH > HCHO$ (SCHNEIDER *et al.* 1983).

Es war das Ziel dieser Arbeiten, die Abhängigkeiten der Effektorwirkung der zugesetzten C_1-Verbindungen von der Wachstumsrate des Bakterienstammes GB 25 zu untersuchen und modellmäßig zu erfassen.

Material und Methoden

Alle Untersuchungen sind mit einem neu isolierten Bakterienstamm der Arbeitsbezeichnung GB 25 ZIMET B 502 durchgeführt worden (WENDLANDT *et al.*, in Vorbereitung).

Die experimentellen Versuche wurden in einem 15-l-Rührfermentor kontinuierlich unter Chemostatbedingungen durch Variation des Methan/Sauerstoff-Verhältnisses bei sonst konstanter Gesamtgasmenge und Rührereintragsleistung durchgeführt.

Als konstante Prozeßbedingungen wurden eingestellt: pH-Wert 5,1, Fermentationstemperatur 38 °C. Die Nährsalze wurden so dosiert, daß weder Limitationen noch Inhibierungen auftreten konnten (WENDLANDT *et al.*, in Vorbereitung).

Die C_1-Verbindungen wurden mit dem Nährmedium in den Fermentor dosiert, entsprechend einer Menge pro 1 g gebildete Biomasse von: 0,023 g CH_3OH, 0,022 g HCHO, 0,018 g HCOOH und 0,022 g CO_2 (als Bicarbonatäquivalente). Die Durchflußraten wurden im Bereich von $D = 0{,}067$ bis $0{,}167\ h^{-1}$ variiert.

Modellansätze: Die Abhängigkeit des Ausbeutekoeffizienten Y von der Wachstumsrate der Mikroorganismen kann nach der bekannten linearen Beziehung von PIRT (1965) bzw. SCHULZE und LIPE (1964) beschrieben werden:

$$\frac{1}{Y} = \frac{1}{Y^\infty} + \frac{m}{Y} \tag{1}$$

Durch Auflösung der Gleichung 1 nach Y erhält man eine hyperbolische Funktion der Form:

$$Y = Y^\infty \frac{\mu}{\mu + mY^\infty} \tag{2}$$

bzw. nach dem Stationäritätsprinzip mit $\mu = D$ gilt:

$$Y = Y^\infty \frac{D}{D + mY^\infty} \tag{3}$$

Die beiden Koeffizienten m und Y^∞ in Gleichung 3 können als systemspezifische Konstanten angesehen werden, so daß auch das Produkt $m \times Y^\infty$ zu der Systemkonstante K_D zusammengezogen werden kann. Der Betrag von K_D stellt diejenige Wachstumsrate dar, bei welcher der Wert $Y^\infty/2$ erreicht wird. So folgt aus (3) Gleichung 4:

$$Y = Y^\infty \frac{D}{D + K_D} \tag{4}$$

Gleichung 4 besitzt eine Analogie zur Adsorptionsisotherme für heterogenkatalytische Reaktionen und auch zum Wachstumsmodell nach MONOD.

Für reale Systeme ist Y^∞ eine fiktive Größe, da die wahre maximale Ausbeute Y^{max} noch vor Erreichen der maximalen Wachstumsrate bzw. auch der kritischen Durchflußrate erhalten wird. Definitionsgemäß wird die zum Y^{max}-Wert gehörende Durchflußrate als optimale Durchflußrate D_{opt} bezeichnet, so daß aus Gleichung 4 folgt:

$$Y^{max} = Y^\infty \frac{D_{opt}}{D_{opt} + K_D} \tag{5}$$

Im Falle der kontinuierlichen Kultivierung des Bakterienstammes GB 25 beträgt $D_{opt} \simeq 0{,}95\, D_{crit}$.

Des weiteren wurde ein stöchiometrischer Quotient $\alpha^{O_2}_{CH_4}$ gebildet, welcher das Verhältnis zwischen Sauerstoff- und Methanverbrauch pro gebildete Biomasseeinheit widerspiegelt. Dieser Quotient wurde aus

$$\alpha^{O_2}_{CH_4} = Y_S/Y_{O_2} \tag{6}$$

kalkuliert.

Ergebnisse

CH_4-Limitation

Zusätze von sauerstoffhaltigen C_1-Verbindungen, die zugleich auch C_1-Intermediate im C_1-Metabolismus sind, erhöhen sowohl die maximale Wachstumsrate (kritische Durchflußrate) als auch die Ausbeutekoeffizienten bezüglich Methan und Sauerstoff. Die erhaltenen experimentellen Daten lassen sich sehr gut durch die Modelle der Gleichung 1 bzw. 3 beschreiben (vgl. Abb. 1 und 2).

Die nach Gleichung 1 ermittelten Koeffizienten Y^∞ und m sowie die experimentell bestimmten kritischen Durchflußraten D_{crit}, sowie die für D_{opt} berechneten Y_{max}-Werte sind zusammenfassend für die einzelnen C_1-Verbindungen in den Tabellen 1 (CH_4-bezogene Koeffizienten) und 2 (O_2-bezogene Koeffizienten) aufgeführt.

Tabelle 1

Kalkulierte methanbezogene Koeffizienten der Gleichung 1 sowie ermittelte kritische Durchflußraten ohne und mit zugeführten sauerstoffhaltigen C_1-Verbindungen beim C-limitierten Wachstum des Bakterienstammes GB 25

Zusätze	Y_S^∞ (g BTS/ g CH_4)	m_S (g CH_4/ g BTS · h)	D_{crit} (h^{-1})	Y_S^{max} (g BTS/ g CH_4)
CH_3OH	1,15	0,048	0,25	0,94
HCHO	0,96	0,039	0,21	0,81
HCOOH	1,03	0,042	0,22	0,85
CO_2	1,08	0,045	0,24	0,89
—	0,86	0,034	0,20	0,76

Tabelle 2

Sauerstoffbezogene Wachstumskoeffizienten nach Gleichung 1 ohne und mit zugeführten sauerstoffhaltigen C_1-Verbindungen beim C-limitierten Wachstum des Bakterienstammes GB 25 (D_{crit}-Werte vgl. Tab. 1)

Zusätze	$Y_{O_2}^\infty$ (g BTS/g O_2)	m_{O_2} (g O_2/g BTS · h)	$Y_{O_2}^{max}$ (g BTS/g O_2)
CH_3OH	0,42	0,166	0,32
HCHO	0,34	0.147	0,27
HCOOH	0,34	0,147	0,27
CO_2	0,38	0,157	0,30
—	0,27	0,123	0,23

Man erkennt aus der grafischen Darstellung der Abbildung 1, daß oberhalb eines Schnittpunktes der Kurvenschar die effektorische Wirkung der zugeführten C_1-Verbindungen progressiv anwächst und praktisch mit Erreichen der optimalen Durchflußrate endet. Dieser Schnittpunkt kann als sogenannter isoeffektorischer Punkt bezeichnet werden, welcher im Falle der Y_s-Werte durch $D_{iso} = 0{,}04\ h^{-1}$ und $Y_S^{iso} = 0{,}50$ g BTS/g CH_4 und für die Y_{O_2}-Werte durch $D_{iso} = 0{,}033\ h^{-1}$ und $Y_{O_2}^{iso} = 0{,}14$ g BTS/g O_2 bestimmt wird.

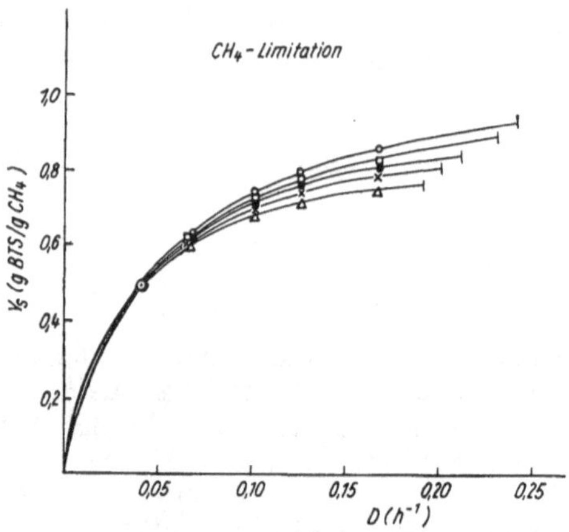

Abb. 1. Einfluß zugesetzter C_1-Verbindungen auf die Ausbeute (Y_S) in Abhängigkeit von der Durchflußrate bei C-Limitation. (○) CH_3OH, (□) CO_2, (●) HCOOH, (×) HCHO, (△) ohne Zusatz; D_{opt}; (—) Modellkurve nach Gleichung 3 und Daten der Tabelle 1

Unterhalb der isoeffektorischen Durchflußrate ($D < D_{iso}$) besitzen die zugeführten C_1-Verbindungen eine schwach ausgeprägte inhibierende Wirkung, die sich in der Verringerung der Y-Werte im Vergleich zur Fermentation ohne Zusätze ausdrückt.

Prinzipiell steigt die Effektorwirkung bei hoher Durchflußrate in der Reihenfolge:

$$CH_3OH > CO_2 > HCOOH > HCHO$$

In dieser Reihenfolge nehmen auch unter vergleichbaren Fermentationsbedingungen die Werte für die Systemkoeffizienten Y^{max}, Y^{∞}, m, D_{opt} und K_D ab.

Sauerstofflimitation

Unter den Bedingungen einer chemostatischen O_2-Limitation werden analoge Beziehungen wie für die C-limitierten Fermentationen erhalten (Tab. 3 und 4).

So fällt auch hier die Effektorwirkung in der Reihe:

$$CH_3OH > CO_2 > HCOOH > HCHO$$

Gegenüber dem C-limitierten Prozeß sind die Y_s, Y_s^{max}- und m_s-Werte niedriger, dagegen die entsprechenden $Y_{O_2}^{max}$- und m_{O_2}-Werte sowie die kritische Durchflußrate höher. Auch beim O_2-limitierten Prozeß treten isoeffektorische Punkte bei $D_{iso} = 0{,}036\ h^{-1}$ und $Y_s = 0{,}5$ g BTS/g CH_4 bzw. $Y_{O_2} = 0{,}13$ g BTS/g O_2 auf (vgl. hierzu

Tabelle 3

Methanbezogene Wachstumskoeffizienten nach Gleichung 1 mit und ohne Zuführung sauerstoffhaltiger C_1-Verbindungen unter Sauerstofflimitation

C_1-Verbindung	Y_s^{∞} (g BTS/ g CH_4)	m_s (g CH_4/ g BTS · h)	D_{crit} (h^{-1})	Y_s^{max} (g BTS/ g CH_4)
CH_3OH	1,02	0,036	0,26	0,89
HCHO	0,87	0,031	0,27	0,79
HCOOH	0,91	0,032	0,26	0,81
CO_2	0,96	0,033	0,26	0,85
—	0,80	0,026	0,20	0,72

Abb. 2). Die höchsten Ausbeutesteigerungen werden unter den gewählten Prozeßbedingungen bei Methanolzusätzen von 0,022 g CH_3OH/g BTS erreicht. Sie betragen 23% für die Y_s^{max}-Werte und 40% für die $Y_{O_2}^{max}$-Werte gegenüber den Fermentationen ohne Zuführung von Methanol.

Tabelle 4
Sauerstoffbezogene Wachstumskoeffizienten nach Gleichung 1 mit und ohne Zuführung sauerstoffhaltiger C_1-Verbindung unter O_2-Limitation

C_1-Verbindung	$Y_{O_2}^{\infty}$ (g BTS/g O_2)	m_{O_2} (g O_2/g BTS · h)	$Y_{O_2}^{max}$ (g BTS/g O_2)
CH_3OH	0,47	0,202	0,34
HCHO	0,34	0,173	0,28
HCOOH	0,38	0,191	0,29
CO_2	0,39	0,210	0,30
—	0,29	0,147	0,24

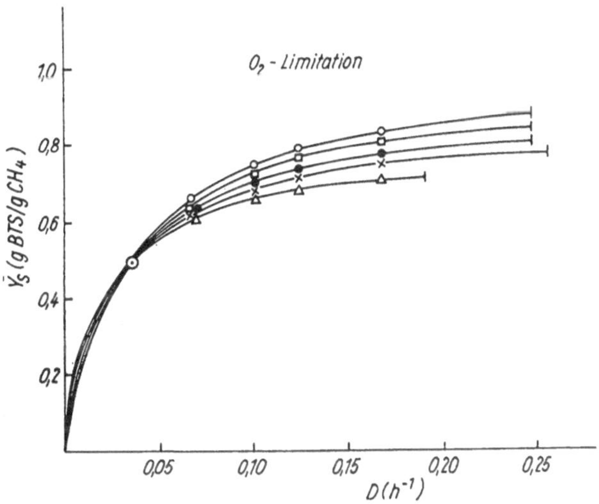

Abb. 2. Einfluß zugesetzter C_1-Verbindungen auf die Ausbeute (Y_S) in Abhängigkeit von der Durchflußrate bei O_2-Limitation. (○) CH_3OH, (□) CO_2, (●) HCOOH, (×) HCHO, (△) ohne Zusatz; (⊣) $D_{opt.}$; (—) Modellkurve nach Gleichung 3 und Daten der Tabelle 3

Stöchiometrie der Biomassesynthese

Die Zusätze der sauerstoffhaltigen C_1-Verbindungen beeinflussen die Bilanz der mikrobiologischen Methanutilisation.

Die Zusammensetzung der Biomasse ändert sich unter dem Einfluß der C_1-Zusätze nicht nachweisbar. Sie beträgt $C_{0,04}\,H_{0,07}\,O_{0,02}\,N_{0,01}$.

Aus der Massenbilanz ist zu erkennen, daß sich durch diese Zusätze die spezifischen Mengen an Endprodukten (CO_2, H_2O) verringern und dadurch der spezifische Verbrauch für Sauerstoff und Methan herabgesetzt wird (Tab. 5).

Das Verhältnis der utilisierten spezifischen Mengen an Methan und Sauerstoff ist nicht konstant, es ist u. a. abhängig von der Art der Zusätze der C_1-Intermediate, der Wachstumsgeschwindigkeit der Mikroorganismen und den ausgewählten Limitationsbedingungen (vgl. Abb. 3 und 4):

Tabelle 5
Massenbilanz der Zellvermehrung der Bakterien GB 25 unter den Bedingungen $D = 0{,}167\ h^{-1}$, O_2-Limitation und Zusätzen an C_1-Verbindungen ($CH_3OH = 0{,}023$ g/g BTS; $HCHO = 0{,}022$ g/g BTS; $HCOOH = 0{,}018$ g/g BTS und CO_2 (als Bicarbonat) $= 0{,}022$ g/g BTS) sowie 0,12 g NH_3/g BTS)

Zusatz an C_1-Verbindungen	
ohne	1,40 g CH_4 + 4,32 g O_2 → 1 g BTS + 2,03 g CO_2 + 2,48 g H_2O
HCHO	1,34 g CH_4 + 3,95 g O_2 → 1 g BTS + 1,88 g CO_2 + 2,35 g H_2O
HCOOH	1,29 g CH_4 + 3,74 g O_2 → 1 g BTS + 1,74 g CO_2 + 2,25 g H_2O
CO_2	1,24 g CH_4 + 3,68 g O_2 → 1 g BTS + 1,61 g CO_2 + 2,12 g H_2O
CH_3OH	1,20 g CH_4 + 3,28 g O_2 → 1 g BTS + 1,50 g CO_2 + 20,7 g H_2O

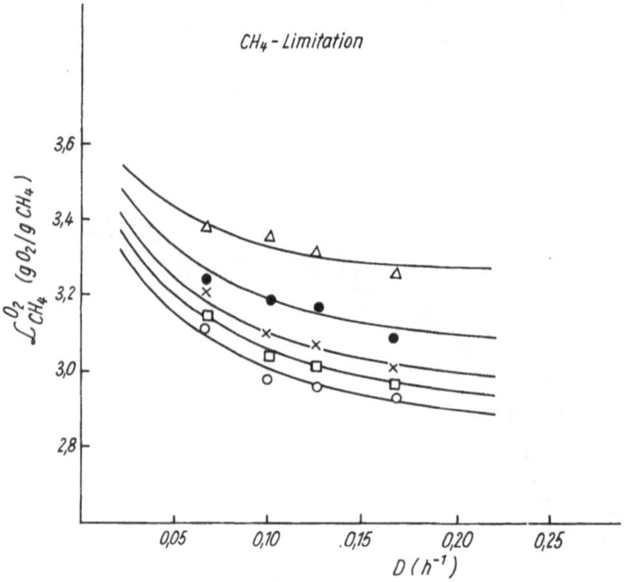

Abb. 3. Abhängigkeit des stöchiometrischen Quotienten $\alpha_{CH_4}^{O_2}$ von der Durchflußrate und der Art der zugesetzten C_1-Verbindungen unter CH_4-Limitation (Symbole für C_1-Verbindungen wie in Abb. 1 und 2)

$$\alpha_{CH_4}^{O_2} = f\ (\mu,\ C_1\text{-Zusätze, Limitationsart})$$

Der stöchiometrische Quotient $\alpha_{CH_4}^{O_2}$ kann als ein Maß für die Effizienz der Biomassesynthese angesehen werden. Je kleiner dieser Quotient ist, desto größer ist die Zellausbeute und umso geringer ist die CO_2- und Wasserbildung.

Aus Gleichung 6 resultiert für den Fall einer unendlich hohen Wachstumsrate:

$$(\alpha_{CH_4}^{O_2})^{\infty} = Y_s^{\infty} Y_{O_2}^{\infty} \tag{7}$$

Der Zusammenhang zwischen Gleichung 6 und 7 kann im untersuchten Bereich der Durchflußrate von $D = 0{,}167 \dots 0{,}067\ h^{-1}$ in erster Näherung durch eine Geradengleichung beschrieben werden (Abb. 5):

$$\alpha_{CH_4}^{O_2} = (\alpha_{CH_4}^{O_2})^{\infty} + K_{O/C} \cdot D^{-1} \tag{8}$$

Die Kurvenschar in Abbildung 5 für die CH_4-Limitation verläuft nahezu parallel, wie auch aus den $K_{O/C}$-Werten in Tabelle 6 zu erkennen ist. Hingegen werden bei

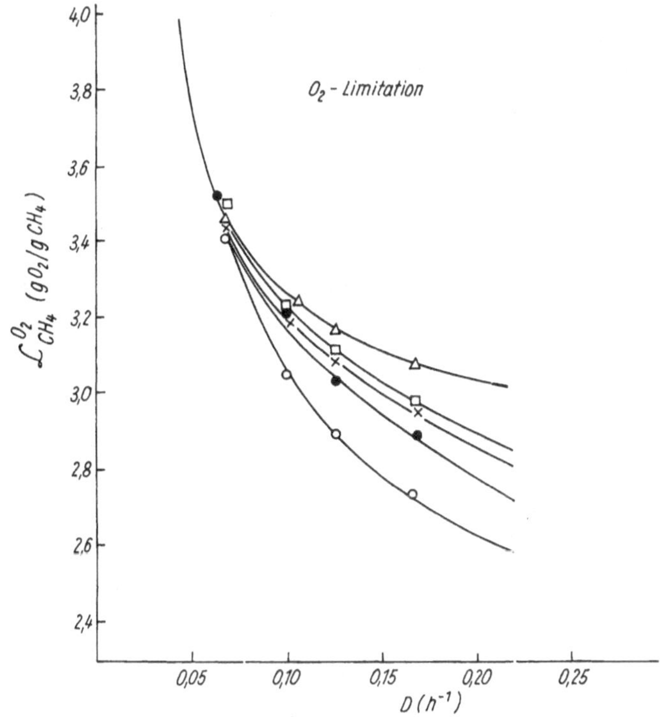

Abb. 4. Abhängigkeit des stöchiometrischen Quotienten $\alpha^{O_2}_{CH_4}$ von der Durchflußrate und der Art der zugesetzten C_1-Verbindungen (Symbole siehe Abb. 1 bzw. 2 unter O_2-Limitation)

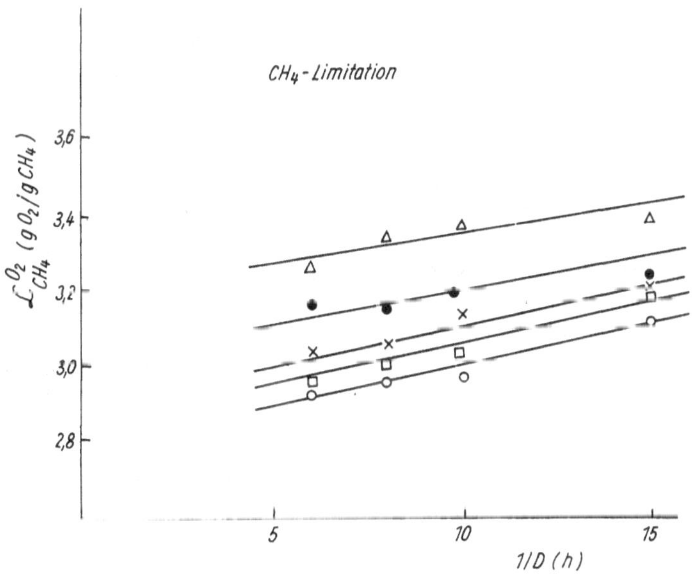

Abb. 5. Lineare Korrelation zwischen dem $\alpha^{O_2}_{CH_4}$-Wert und der reziproken Durchflußrate im Bereich zwischen 5 und 15 h unter CH_4-Limitation (Symbole für C_1-Verbindungen wie in Abb. 1 und 2)

Sauerstofflimitation sowohl der Koeffizient $K_{O/C}$ als auch die $(\alpha^{O_2}_{CH_4})^{\infty}$-Werte durch die C_1-Zusätze beeinflußt. Der Koeffizient $K_{O/C}$ für den O_2-limitierten Prozeß ist etwa 2—3mal höher als für den C-limitierten Prozeß. Dagegen werden im O_2-limitierten Prozeß geringere stöchiometrische Quotienten $(\alpha^{O_2}_{CH_4})^{\infty}$ erhalten. Das praktisch erreichbare Minimum der stöchiometrischen Quotienten $\alpha^{O_2}_{CH_4}$ wird bei den jeweiligen optimalen Durchflußraten erreicht; sie betragen 2,6 ... 3,0 g O_2/g CH_4 (O_2-Limitation) und 2,9—3,3 g O_2/g CH_4 (CH_4-Limitation).

Tabelle 6
Stöchiometrische Quotienten $\alpha^{O_2}_{CH_4}$ für die methan- und sauerstofflimitierte Fermentation. Daten kalkuliert nach Gleichung 8

Zusatz an C_1-Verbindung	CH_4-Limitation		O_2-Limitation	
	$\alpha^{O_2}_{CH_4}$ (g O_2/g CH_4)	$K_{O\|C}$ (g O_2/g $CH_4 \cdot$ h)	$\alpha^{C_2}_{CH_4}$ (g O_2/g CH_4)	$K_{O\|C}$ (g O_2/g $CH_4 \cdot$ h)
CH_3OH	2,76	0,022	2,30	0,075
HCHO	2,88	0,022	2,63	0,055
HCOOH	3,03	0,020	2,51	0,070
CO_2	2,84	0,021	2,63	0,060
~	3,21	0,020	2,84	0,041

Diskussion

Frühere Untersuchungen zeigten, daß durch Zusätze von sauerstoffhaltigen C_1-Verbindungen eine stimulierende oder inhibierende Wirkung bei der kontinuierlichen Kultivierung des Bakterienstammes GB 25 erzielt werden kann (Schneider *et al.* 1983). Methanol, Formaldehyd, Ameisensäure und Kohlendioxid beeinflussen sowohl die Ausbeute als auch die maximale Wachstumsrate (Abb. 1). Eine maximale Stimulierung des Wachstums wird bei Zusätzen der C_1-Verbindungen von *ca.* 2×10^{-2} g C_1-Verbindung/g BTS erreicht. Unter diesen Bedingungen werden Ausbeutesteigerungen von 10—20% erreicht, die maximale Wachstumsrate wird bis zu 25% erhöht. Dabei beträgt der Anteil der C_1-Verbindungen an der Biomassesynthese $< 2\%$, so daß sie lediglich katalytisch wirksam werden.

Die Effektorwirkung der C_1-Zusätze steigt mit zunehmender Wachstumsgeschwindigkeit der Bakterien an. Vom Methanol geht die größte Wirkung aus, gefolgt von CO_2, Formiat und Formaldehyd. Diese Wirkung bezieht sich auf die erreichbaren maximalen Ausbeuten Y^{max}. Die effektorische Wirkung spiegelt sich in der Größe der Erhaltungskoeffizienten m, der K_D-Werte und des stöchiometrischen Quotienten des spezifischen O_2- und CH_4-Verbrauchs $\alpha^{O_2}_{CH_4}$ wider.

Die Ergebnisse weisen insgesamt darauf hin, daß die maximale Wachstumsrate und die Ausbeutekoeffizienten durch zentrale Regelgrößen bestimmt werden, welche über das Verhältnis zwischen oxidativem Formaldehyd-Abbau und C_1-Assimilation (Transformylierung an Glycin) entscheiden.

Im Serin-Metabolismus der methylotrophen Bakterien spielt nach verschiedenen Untersuchungen die NADH-Konzentration eine entscheidende Rolle (Babel u. Mothes 1980, Higgins *et al.* 1981, Newaz u. Hersh 1975, Harder u. Attwood 1978).

Für die Aufnahme mindestens einer Regelgröße spricht das Auftreten von isoeffektorischen Punkten. Außerdem zeigten spektroskopische Untersuchungen, daß bei Zusätzen von sauerstoffhaltigen C_1-Verbindungen eine Korrelation zwischen der NADH-Konzentration, der maximalen Wachstumsrate und der Zellausbeute existiert (Wend-

LANDT *et al.*, in Vorbereitung). Damit erhöhen die C_1-Zusätze den Pool an Energieäquivalenten, z. B. in Form von NADH in der Bakterienzelle. Es ist anzunehmen, daß die NADH-Konzentration die Effizienz der Biomassesynthese bzw. das Verhältnis zwischen assimilatorischer bzw. oxidativer Verwertung des Formaldehyds zu steuern scheint.

Symbolverzeichnis

BTS Trockensubstanz für Bakterien
D Durchflußrate (h^{-1})
D_{crit} Kritische Durchflußrate für maximale Wachstumsrate bei chemostatischer Fermentation (h^{-1})
D_{opt} Optimale Durchflußrate für $Y^{max}(h^{-1})$
K_D Geschwindigkeitskoeffizient für $Y^{\infty}/2$ (h^{-1})
$K_{O/C}$ Systemspezifische Konstante ($g\ O_2/g\ CH_4 \cdot h$)
m Koeffizient für Substrataufnahmerate (Erhaltungskoeffizient) (g S/g BTS · h)
S C-Substrat (Methan)
Y Ausbeutekoeffizient (g BTS/g S bzw. g BTS/g O_2)
Y^{max} Maximaler Ausbeutekoeffizient für D = optimal
Y^{∞} Ausbeutekoeffizient für $D = \infty$
$\alpha_{CH_4}^{O_2}$ Stöchiometrischer Umsatzkoeffizient ($g\ O_2/g\ CH_4$)
μ Spezifische Wachstumsrate (h^{-1})

Literatur

BABEL, W. und MOTHES, G., 1978. Dissimilatorische Sequenzen in methylotrophen Bakterien. Z. Allg. Mikrobiol., **18**, 17—26.

BABEL, W. und MOTHES, G., 1980. Rolle der Formiat-Dehydrogenase in Serinweg-Bakterien. Z. Allg. Mikrobiol., **20**, 167—175.

BROWN, L. R., STRAWINSKI, R. J. and MC.CLESKY, C. S., 1964. The isolation and characterisation of *Methanomonas methanooxidans*. Canad. J. Microbiol., **10**, 791.

HARDER, W. and ATTWOOD, M. M., 1978. Biology, physiology and biochemistry of Hyphomicrobia. Adv. Microb. Physiol., **17**, 303—359.

HIGGINS, I. J., BEST, D. J., HAMMOND, R. C. and SCOTT, D., 1981. Methane-oxidizing microorganisms. Microbial Rev., **45**, 556—590.

JOHNSON, P. A. and QUAYLE, J. R., 1964. Oxidation of methanol, formaldehyd and formiate by methanol-grown *Pseudomonas* AM 1. Biochem. J., **93**, 281.

NEWAZ, S. S. and HERSH, L. B., 1975. Reduced nicotinamide adenine dinucleotide-activated phosphoenolpyruvate carboxylase in *Pseudomonas* MA: potential regulation between carbon assimilation and energy production. J. Bacteriol., **124**, 825—833.

PAPDUTSAKIS, E., HIRT, W. and LIM, H. C., 1981. Bacterial utilization of pure and mixed C_1 compounds for single cell protein production. Biotechnol. Bioeng., **23**, 235.

PILAT, P. and PROKOP, A., 1975. The effect of methanol, formaldehyde and formic acid on growth of *Candida boidinii*. Biotechnol. Bioeng., **17**, 1717.

PIRT, S. J., 1965. The maintenance energy of bacteria in growing cultures. Proc. Royal Soc. B, **163**, 224—231.

QUAYLE, J. R., 1972. The metabolism of one-carbon compounds by microorganisms. Adv. Microb. Physiol., **7**, 119.

RICHTER, K., ROGGE, G. und WENDLANDT, K.-D., 1974. Verfahren zur Kultivierung von Mikroorganismen. Patent DD-PS 104097

SCHNEIDER, J. D., WENDLANDT, K.-D., BRÜHL, E. und MIRSCHEL, G., 1983. Einfluß sauerstoffhaltiger C_1-Verbindungen auf das Wachstum methanassimilierender Bakterien. Z. Allg. Mikrobiol., **23**, 33.

Schulze, K. L. and Lipe, R. S., 1964. Relationship between substrate concentration, growth rate and respiration rate of *Escherichia coli* in continuous culture. Arch. Mikrobiol., **48**, 1–20.

Wendlandt, K.-D. und Brühl, E., 1980. Phänotypische Optimierung der bakteriellen Methankonvertierung mittels Carbonat/Kohlendioxid. 2. Symposium der sozialistischen Länder über Biotechnologie, Dez. 1980 Leipzig, Poster.

Wendlandt, K.-D., Schneider, J. D., Mirschel, G., Geier, W., Bayer, C. and Nagel, B., 1982. Investigation of the NADH signal of methanotrophic bacteria. VIII. Leipziger Biotechnologie-Symposium, Leipzig, Sept. 1982, Poster.

Anschrift: Dr. K.-D. Wendlandt
Institut für technische Chemie
der Akademie der Wissenschaften der DDR
DDR 7050 Leipzig, Permoser Str. 15

Zeitschrift für Allgemeine Mikrobiologie	23	1983	4	269–273

Kurze Originalmitteilung

(Akademie der Wissenschaften der DDR, Forschungszentrum für Molekularbiologie und Medizin, Zentralinstitut für Mikrobiologie und experimentelle Therapie, Jena, Direktor: Prof. Dr. habil. U. TAUBENECK und Biologische Laboratorien des VEB Jenapharm, Jena[1])

Sequential processes of phosphate limitation and of phosphate release in streptomycin fermentations

P. J. MÜLLER, A. CHRISTNER[1] and J. H. OZEGOWSKI

(Eingegangen am 20. 7. 1983)

The significance of the sequential processes of phosphate limitation and of phosphate release from medium constituents is demonstrated in technical streptomycin fermentations. The phosphate limitation initiated the streptomycin synthesis as well as the formation of phosphatases and protease. In later periods of the process the phosphate release influences especially the enzyme formation.

In previous kinetical studies on turimycin fermentations (MÜLLER *et al.* 1981, 1982, 1983) the significance of sequential processes of phosphate limitation and of phosphate release from the phosphate reserve of the medium was demonstrated. In this connection the role of phosphate limitation initiating antibiotic biosynthesis is well-known (review: MARTIN and DEMAIN 1980). However, to our knowledge the fundamental significance of the phosphate release from complex phosphate-containing medium constituents was recognized only by ASAI *et al.* (1978) for maridomycin- and by JECHOVÁ *et al.* (1981) for oxytetracyclin fermentations.

In the initial phase of turimycin fermentations phosphate limitation leads to a drastic increase of the activities of dephosphorylating enzymes of different type, *e.g.* phosphatases (MÜLLER *et al.* 1983), thiamine pyrophosphatase (unpublished result) and cAMP phosphodiesterase (RÖMER and MÜLLER 1983). In later phases of the fermentation the release of phosphate, mediated by the extracellular hydrolases, results in a morephasic growth and morephasic behaviour of turimycin and of dephosphorylating enzyme formation, respectively.

To generalize the significance of the sequential processes in technical antibiotic fermentations performed in media with a phosphate reserve, we investigated the kinetical behaviour of a technical streptomycin fermentation by measurement of growth, acid-soluble phosphate, streptomycin, phosphatases, ATPase, and protease.

The results are in agreement with our previous statements of the fundamental role of the sequential processes in technical fermentations, not only the well-known effect of phosphate limitation but also the release of phosphate.

The fermentations were performed at 28 °C with *Streptomyces griseus spec.* in a medium containing soya meal, glucose, and inorganic salts. Glucose and ammonia are fed in such a manner, that over the whole process these nutrients are in excess. The methods of sonication, of protease assay, measured by the formation of tyrosine at 280 nm, acid-soluble orthophosphate (P), protein and phosphatase activities, measured by the formation of *p*-nitrophenol, were described by MÜLLER *et al.* (1982). For calculation of the phosphatase activities an absorbance coefficient of *p*-nitrophenol $\varepsilon_{405} = 13.3\ \text{cm}^2/\mu\text{mole}$ at pH 7.2 and $= 18.5\ \text{cm}^2/\mu\text{mole}$ at pH 9.2 was used.

ATPase was estimated according to Hanson and Kennedy (1973) (method D) by the change of absorbance accompanying the removal of NADH, coupled to pyruvate and ADP generation, respectively, in the presence of ATP, PEP, pyruvate kinase and lactate dehydrogenase. Phosphate release was determined every hour by estimation of the acid-soluble phosphate concentration of culture filtrates shaken at 28 °C over a range of five hours.

The total phosphate content of the medium was estimated to be 512 mg/l P · 150 mg/l P were found to be acid-soluble after steam pressure sterilization. With the soya meal, the only complex medium constituent (50 g/l) about 385 mg/l P are transferred into the medium, calculated to contain an approximate value of 9 mg P/ per g soya meal. 46 mg/l P are added to the medium as KH_2PO_4. Accordingly about 150—46 = 104 mg/l P occurring in the soya meal are found to be acid-soluble.

Fig. 1 shows that most of the acid-soluble phosphate is consumed up to the 35th h. Growth continues in a linear manner up to the 62nd h. It is now assumed that a certain part of the phosphate reserve of 512—150 = 362 mg/l P; is released by the action of extracellular enzymes. Especially the relatively weak second growth period (60—70thh) and possibly also a third growth phase (80—90th h), superimposed by cell-lytic processes, are assumed to result from phosphate release and not from irregularities of growth measurement. There are some hints supporting this assumption.

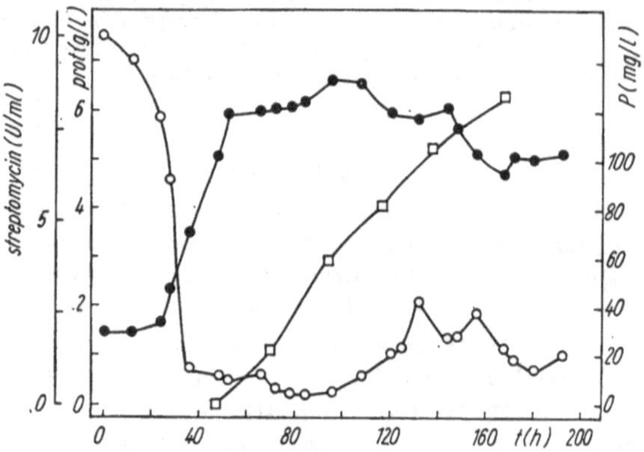

Fig. 1. Kinetics of growth measured as intracellular protein (●), acid-soluble orthophosphate (○), and streptomycin (□) of a fermentation with *S. griseus spec.*

Fig. 2 shows the ability of cultural filtrates of different age to release acid-soluble phosphate. The kinetical behaviour, characterized by two maxima, probably does not reflect the concerted enzymatic activity releasing phosphate but the occurrence of dissolved hydrolyzable nutrients in the filtrates. Accordingly, the second growth period is restricted by a transient deficiency of such compounds in the filtrate. Evidence of a true morephasic growth in connection with phosphate release is supported also by the kinetics of the intracellular phosphatases, especially of alkaline phosphatase (Fig. 3, 4).

The maximum values of alkaline phosphatase are observed in the stationary phases of growth and are discussed to be due to a repression of phosphatases by phosphate. This effect of phosphate on phosphatases from *Streptomyces* has been observed by different authors (Majumdar and Majumdar 1971, Toropová *et al.* 1973, Jeloková *et al.* 1974, Madry *et al.* 1979, Jechová *et al.* 1981).

The possibility of assessing the state of phosphate nutrition by the degree of expression of alkaline phosphatase has already been proposed, for example by Fitzgerald (1969) for algae, and the correlation to phosphate release and resulting growth, respectively, was demonstrated by our investigations on turimycin fermentations. Also the

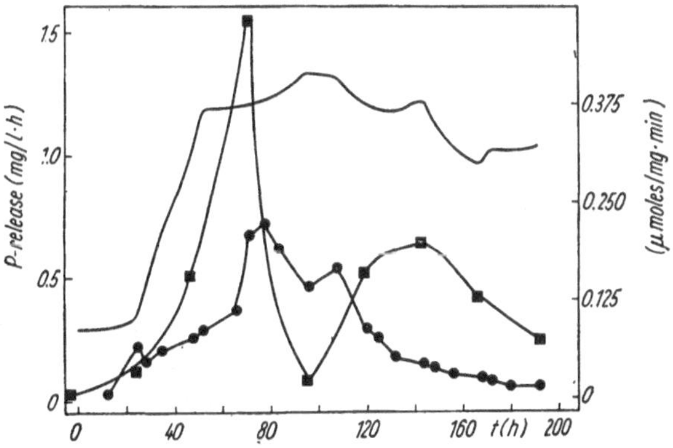

Fig. 2. Kinetics of the specific intracellular ATPase activity (●) and of the phosphate-releasing activity of culture filtrates (■) of the fermentation shown in Fig. 1. The slope of growth is outlined for comparison here and in the following figures

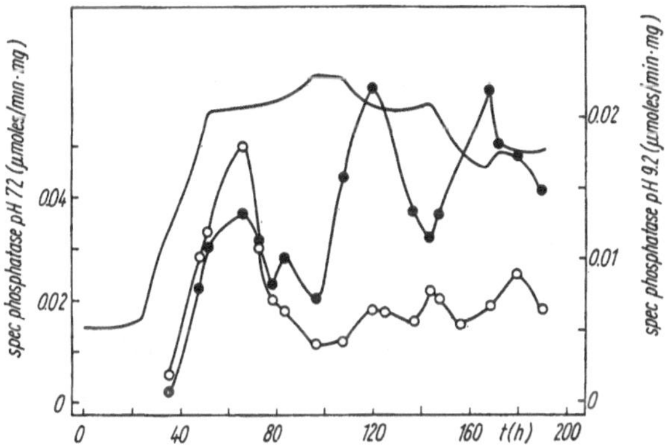

Fig. 3. Kinetics of the intracellular specific activities of phosphatase at pH 7.2 (○) and at pH 9.2 (●)

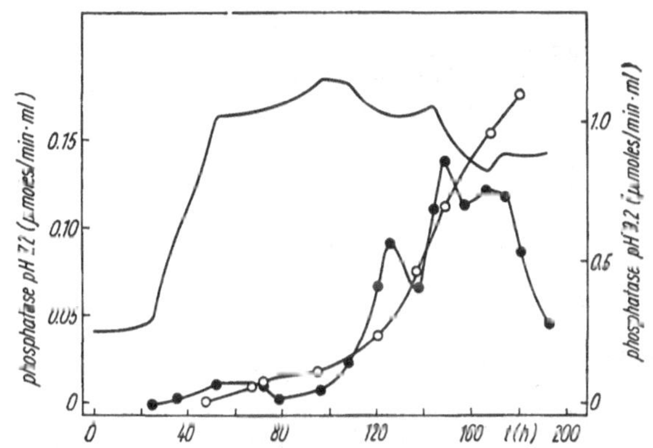

Fig. 4. Kineties of extracellular activities of phosphatase at pH 7.2 (○) and at pH 9.2 (●)

ATPase activity increases especially during the first stationary phase and transiently during the second stationary phase indicating a growth- and phosphate-regulated expression of synthesis (Fig. 2).

Previously, as a consequence of phosphate limitation a partial uncoupling of respiration has been observed in turimycin fermentations. The high ATPase activity in streptomycin fermentations at phosphate limitation may be discussed in the same direction.

In accordance with our assumption of a derepression of the enzyme synthesis at phosphate limitation in secondary product fermentations (Müller *et al.* 1983) besides the phosphatases and ATPase also the protease and, therefore, also the extracellular protein concentration increases in a first approximation parallel with streptomycin (Fig. 5).

Thus, phosphate limitation may be regarded as a general signal to initiate protein and secondary product formation, respectively. The fermentation kinetics are also modified by the process of phosphate release influencing parameters which are under regulatory control by phosphate.

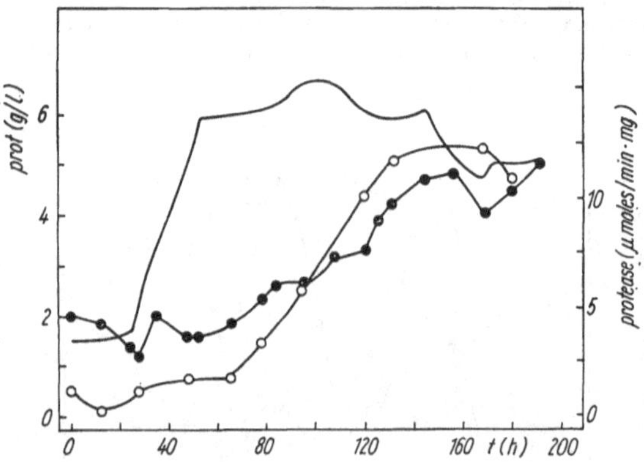

Fig. 5. Kinetics of extracellular protein (●) and extracellular proteolytic activity (○)

References

Asai, T., Sawada, H., Yamaguchi, T., Higashide, E., Kono, T. and Uchida, M., 1978. Kinetic study of the production of maridomycin by *Streptomyces*. J. Ferment. Technol., **121**, 91–96.

Fitzgerald, G. P., 1969. Field and laboratory evaluations of bioassays for nitrogen and phosphorus with algae and aquatic weeds. Limnol. Oceanogr., **14**, 206–212.

Hanson, L. H. and Kennedy, E. P., 1973. Energy-transducing adenosine triphosphatase from *Escherichia coli*. J. Bacteriol., **114**, 772–781.

Hošťálek, Z., 1980. Catabolite regulation of antibiotic biosynthesis. Folia microbiol., **25**, 445 to 450.

Jechová, V., Čurdová, E. and Hošťálek, Z., 1981. The role of phosphatases in the biosynthesis of chlortetracycline in *Streptomyces aureofaciens*. Poster abstracts of the FEMS Symposium on "Overproduction of Microbial Products", 9.–14. August 1981, Hradec Králove, Czechoslovakia, P 50.

Jeloková, J., Zelinková, E., and Zelinka, J., 1974. Activity of nucleolytic enzymes in mycelium of *Streptomyces aureofaciens* during fermentation. Biologia (Bratislava), **29**, 207.

Madry, N., Sprinkmeyer, R. and Pape, H., 1979. Regulation of tylosin synthesis in *Streptomyces*: Effects of glucose and inorganic phosphate. Eur. J. Appl. Microbial., **7**, 365–370

Majumdar, M. K. and Majumdar, S. K., 1971. Relationship between alkaline phosphatase and neomycin formation in *Strep. fradiae*. Biochem. J., **122**, 397–404.

Martin, J. F. and Demain, A. L., 1980. Control of antibiotic biosynthesis. Microbiol. Rev., **44**, 230–251.

Miller, A. L. and Walker, J. B., 1970. Accumulation of streptomycin phosphate in cultures of streptomycin producers grown on a high phosphate medium. J. Bacteriol., **104**, 8–12.

MÜLLER, P. J., OZEGOWSKI, J. H., BOCKER, H. and FORBERG, W., 1981. Phosphate release in turimycin fermentations. Poster abstracts of the FEMS-Symposium on "Overproduction of Microbial Products", 9.—14. August 1981, Hradec Králové, Czechoslovakia, P 48.

MÜLLER, P. J., HILLIGER, M., MENNER, M. and FORBERG, W., 1982. Phosphate release in turimycin fermentations. Z. Allg. Mikrobiol., **21**, 267—272.

MÜLLER, P. J., OZEGOWSKI, J. H. and BOCKER, H., 1983. Regulation of hydrolase formation and phosphate release in turimycin fermentations. Z. Allg. Mikrobiol., **23**, 173—180.

RÖMER, W. and MÜLLER, P. J., 1983. Parallel regulation of cAMP phosphodiesterase and phosphatase in turimycin fermentations. Z. Allg. Mikrobiol., **23**, 203—206.

TOROPOVA, E. G., EGOROV, N. S. and SUCHKOVA, L. A., 1973. Alkaline phosphatase of *Proactinomyces fructiferi var. risomycini* producing ristomycin. Antibiotiki (Moskov), **18**, 587—590.

Mailing address: Dr. P. J. MÜLLER
Zentralinstitut für Mikrobiologie und
experimentelle Therapie der AdW
DDR-6900 Jena, Beutenbergstr. 11

Zeitschrift für Allgemeine Mikrobiologie	23	1983	4	275–280

Buchbesprechungen

M. B. Bushell and J. H. Slater (Editors), Mixed Culture Fermentations. 175 S., 41 Abb., 20 Tab. London-New York-Toronto-Sydney-San Francisco 1981. Academic Press. $ 25.00.

In der Nahrungsmittel- und Getränkeindustrie, Abfallbeseitigung bis hin zur Energiegewinnung auf mikrobiellem Wege finden sich zahlreiche Beispiele für die Nutzung von Mischkulturen. Obwohl sich gerade auf diesen Gebieten ständig neue Anwendungsmöglichkeiten ergeben, entsprach die Erforschung der Eigenschaften und Mechanismen von Mischkulturen bisher nicht dem wachsenden Interesse.

Das vorliegende Buch enthält acht auf einem Symposium der Society for General Microbiology in London 1980 vorgetragene Übersichtsbeiträge zur Theorie und Praxis der Fermentation von Mischkulturen.

Die ersten beiden Kapitel beinhalten eine dem gegenwärtigen Wissensstand entsprechende Analyse der vielfältigen mikrobiellen Wechselwirkungen in Mischkulturen. Während J. H. Slater im ersten Kapitel anhand von Beispielen eine Klassifizierung mikrobieller Gemeinschaften vorschlägt, behandelt M. J. Bazin die Kinetik der Wechselwirkungen unter Anwendung der mathematischen Modellierung. Die übrigen sechs Kapitel befassen sich mit Entwicklungstendenzen bei der Abfallbeseitigung sowie in der Nahrungsmittel- und Getränkeindustrie. Hervorzuheben ist der Artikel von F. M. Driessen, der anhand eines Beispiels aus der Milchindustrie (Joghurt-Herstellung) in gelungener Weise die Wechselwirkungen zweier Bakteriengattungen (*Streptomyces, Lactobacillus*) aus physiologischer Sicht charakterisiert und mathematisch modelliert. Die Verknüpfung von Theorie und Anwendung in der Praxis kommt hierbei klar zum Ausdruck.

Das Buch stellt die Bedeutung von Mischpopulationen, aber auch die Schwierigkeiten bei ihrem detaillierten Studium dar. Nicht zuletzt durch seine gute Literaturübersicht ist dieser Band ein nützliches Übersichtswerk für Mikrobiologen und besonders für Biotechnologen.

M. Peisker (Jena)

D. Gröger und S. Johne, Mikrobielle Gewinnung von Arzneistoffen. 266 S., 51 Abb., 10 Tab. Berlin 1982. Akademie-Verlag. M 12,50.

Unter dem Begriff „Pharmazeutische Mikrobiologie“ wurde eine straff gegliederte und inhaltlich konzentrierte Darstellung der mikrobiologischen Grundlagen, der Chemie und Biochemie und der Geschichte der Herstellung von Wirkstoffen aus Mikroorganismen erarbeitet. Daß es gelungen ist, das an sich sehr umfangreiche Wissensgebiet in Form eines Taschenbuches abzuhandeln und damit nicht allein Studierenden sondern auch fachnahen und fachfremden Lesern eine wertvolle Zusammenstellung von Informationen in die Hand zu geben, verdient Anerkennung. Während in einschlägigen deutschsprachigen Lehrbüchern der Mikrobiologie und Pharmazie die mikrobielle Produktion von Sekundärmetaboliten zumeist nur kurz aufgeführt wird bzw. nur die chemischen und applikationstechnischen Probleme berücksichtigt werden, ist die vorliegende Monographie bemüht, ein breitgefächertes Spektrum von Informationen, angefangen von einem Überblick über die Produktbildner und ihre Physiologie bis hin zur Biosynthese von Alkaloiden, Vitaminen, Aminosäuren, Enzymen und Mycotoxinen vorzulegen. Die Produktion von Impfstoffen und Biotransformationen durch Mikroorganismen werden ebenfalls berücksichtigt. Kernstück bildet jedoch die Vorstellung von Vertretern der wichtigsten Antibioticagruppen, wobei neben biosynthetisch-strukturellen Aspekten die Methodik der fermentativen Produktion und pharmakologische Eigenschaften einschließlich ihres Wirkmechanismus betont werden. Neuere Entwicklungstendenzen, wie die Produktion von Enzyminhibitoren oder immunomodulatorischen Substanzen wurden ebenso diskutiert wie die Einsatzmoglichkeiten der Gentechnologie in der Wirkstoffgewinnung.

U. Gräfe (Jena)

H. B. Herscowitz, H. T. Holden, J. A. Bellanti and A. Ghaffar (Editors), Manual of Macrophage Methodology — Collection, Characterization, and Function. IX + 531 S., 42 Abb., 26 Tab. New York-Basel 1981. Marcel Dekker. SFr. 185.00.

Forschungen zur physiologischen und pathologischen Rolle von Makrophagen haben nach einer Periode der Prävalenz der Erforschung von Lymphocytenfunktionen in der letzten Dekade eine Renaissance herbeigeführt. Das komplexe System der mononucleären Phagocyten, weltweit in zahllosen Laboratorien untersucht, verlangt die Anwendung einer Vielzahl zumeist diffiziler Methoden, die einer Standardisierung bedürfen, wenn ihre Ergebnisse vergleichbar sein sollen. Dieses Anliegen verfolgt das vorliegende Handbuch mit 58 Einzelbeiträgen von insgesamt 95 Autoren über grundlegende Methoden zur Untersuchung von Makrophagen. Es gliedert sich in 4 Teile. Teil I behandelt die Gewinnung und Anreicherung von Makrophagen aus den verschiedenen anatomischen Lokalisationen. Teil II beinhaltet Methoden zur Charakterisierung mittels morphologischer, cytochemischer Verfahren sowie solcher zur Bestimmung von Membranrezeptoren und Enzymen. Teile III und IV beschreiben Methoden zur Erfassung der Makrophagenfunktionen *in vitro* bzw. *in vivo*. Die Gliederung der einzelnen Artikel ist so weit wie möglich vereinheitlicht mit einer kurzen Einführung, Angaben zu benötigtem Material und Ausrüstung, Testdurchführung, einer Diskussion möglicher Ergebnisse und Alternativverfahren sowie einem Literaturverzeichnis. Hier liegt ein Werk vor, das dem Anfänger den Zugang zur diffizilen immunologischen Methodik erleichtert, aber auch dem Experten wertvolle Hilfe für die praktische Arbeit vermittelt.

R. S. FRITSCH (Jena)

W. HOPPE, W. LOHMANN, H. MARKS und H. ZIEGLER (Herausgeber), Biophysik (2. völlig neu bearbeitete Auflage). XXIV + 980 S., 856 Abb., 95 Tab. Berlin-Heidelberg-New York 1982. Springer. Verlag DM 168,00.

Die von den Herausgebern und 64 Mitarbeitern verbesserte und ergänzte Darstellung der Biophysik geht aus vom Bau der Zelle und von der Struktur der Nucleinsäuren als biologisch wichtigster Gruppe von Makromolekülen. Es folgen Kapitel über Untersuchungsmethoden und über mögliche Wechselwirkungen, über Mechanismen der Energie-Übertragung, über Strahlenbiophysik, biologische Anwendung von Isotopenmethoden, Beziehungen zur Thermodynamik, Bedeutung der Enzyme als Biokatalysatoren, Funktion der Nucleinsäuren, Vorgänge der Self-Assembly, Membran-Modelle und Membran-Funktion, Photobiophysik, Biomechanik, Neurobiophysik, Kybernetik, Evolution, beginnend mit der Selbstorganisation der Materie, — („Leben begann mit dem ersten Strang, der sich selbst replizieren und mutieren konnte") —.

Was mit der „Biophysik" als Ganzes gesehen geboten wird, ist kein schlichtes Lehrbuch sondern eine umfassende Information über biophysikalische Beziehungen, die ein Studium biologischer Grundlagen voraussetzen. Hervorzuheben ist die große Zahl instruktiver Abbildungen, die sich weniger auf das biologische als bekannt vorausgesetzte Objekt als auf modellmäßige und schematische, das Verständnis erleichternde Darstellungen beziehen. Literatur-Hinweise sind an entsprechenden Stellen im Text zusammengestellt. W. SCHWARTZ (Braunschweig)

K. HORIKOSHI and T. AKIBA, Alkalophilic Microorganisms, a New Microbial World. VII + 213 S., 61 Abb., 87 Tab. Tokyo und Berlin-Heidelberg-New York 1982. Japan Scientific Societies Press und Springer Verlag. DM 98,00.

Was die Autoren als eine neue bisher kaum beachtete „alcaline world" bezeichnen und auf neue Enzyme, Antibiotica und sonstige technisch verwertbare Verbindungen untersuchen wollen, sind Bakterien und Pilze, die bei hohen pH-Werten von etwa 10—11 gut wachsen und z. B. in alkalischen Seen mit hohen Salzgehalten, zum Teil aber auch in Böden bei mittleren pH-Bereichen vorkommen und z. B. zu den Gattungen *Bacillus*, *Pseudomonas*, *Corynebacterium* und *Penicillium* gehören und Natrium brauchen. In 7 Kapiteln werden Isolierung und Klassifizierung alkalophiler Mikroorganismen, Wachstumsbedingungen, physiologische Merkmale, Möglichkeiten einer technischen Verwendung behandelt. Das Literaturverzeichnis umfaßt mehr als 3500 Titel.

W. SCHWARTZ (Braunschweig)

A. H. LINTON, Microbes, Man and Animals — the Natural History of Microbial Interactions. XVI + 342 S., 40 Abb., 54 Tab. Chichester-New York-Brisbane-Toronto-Singapore 1982. John Wiley & Sons. £ 19.50.

Dieses Buch greift aus dem großen Gebiet der Medizinischen Mikrobiologie einen Teilaspekt heraus, die Wechselwirkung zwischen Mikroben und Wirtsorganismus. Ein besonderer Vorteil liegt in der Einbeziehung der human- und veterinärmedizinischen Aspekte und in der Berücksichtigung von Ergebnissen, die erst kurze Zeit vor dem Erscheinen des Werkes bekannt wurden. Der erste

Teil ist vorwiegend den Eigenschaften der Erreger gewidmet, soweit sie für den Infektions- und Ausbreitungsprozeß wesentlich sind und die durch sie hervorgerufenen Krankheitsprozesse und Abwehrvorgänge seitens des Wirts. Der gesunde Wirtsorganismus wird ebenso berücksichtigt wie der kranke, abgehandelt werden die Mechanismen der Pathogenität, Exo- und Endotoxine. Der zweite Teil ist den allgemeinen Grundlagen der Epidemiologie und der Bekämpfung von Infektionen durch Immunisierung, Chemotherapie, Expositionsprophylaxe sowie den Hospitalinfektionen gewidmet. Ein dritter Teil schließlich enthält ausgewählte epidemiologische Fragestellungen: Zoonosen, Lebensmittelvergiftungen, Wasserepidemien, aber auch ganz spezielle Beiträge wie Gonorrhoe, Brucellosen, Pilzinfektionen. An Hand dieser Beispiele werden die epidemiologischen Aussagen verallgemeinert, können aber ihrerseits durch ihre detaillierte Darstellung tiefer in die Materie eindringen. — Das mit illustrativen Schemata und instruktiven Abbildungen versehene Buch kann kein Lehrbuch der Medizinischen Mikrobiologie ersetzen, es ist aber eine vorzügliche Einführung in die Problematik der Gast-Wirt-Beziehungen, die nicht nur den Anfänger umfassend informiert sondern auch dem Fortgeschrittenen Anregungen gibt. W. Köhler (Jena)

E. Müller und W. Loeffler, Mykologie, Grundriß für Naturwissenschaftler und Mediziner (4. überarbeitete und erweiterte Auflage). 366 S., 188 Abb., 31 Tab. Stuttgart-New York 1982. G. Thieme. DM 34,00.

Die vierte überarbeitete und um (nur) 26 Seiten erweiterte Auflage des längst zum unentbehrlichen Standard avancierten „Müller-Loeffler" ist sehr zu begrüßen. Durch Umstellungen sowie überall eingefügte relevante Ergänzungen hat das Buch sehr gewonnen. Dies gilt besonders für den Abschnitt „Angewandte Mykologie", bei dem den heutigen Interessenschwerpunkten Rechnung getragen worden ist (z. B. Sekundärmetabolite, Antibiotica, Mykotoxine, Recycling-Verfahren unter Mitwirkung von Pilzen, Humanmykosen u. a. m.). Die Zusammenfassung der Schleimpilze und niederen Pilze mit beweglichen Stadien unter Einschluß der Algen und Protozoen zum Reich der Protisten (Protoctista) und der Zygomyceten, Ascomyceten und Basisdiomyceten (inkl. Flechtenpilze und Deuteromyceten) zum eigenen Reich der Fungi ist zunächst ungewohnt, erscheint aber plausibel. Insgesamt hat sich das Taschenbuch unter weitgehender Erhaltung des Umfanges zu einem handfesten Lehrbuch „gemausert", das man jedem empfehlen kann, der einen Einstieg in die verschiedensten Aspekte der Mykologie sucht. Lediglich die Fülle der Begriffe mag dem Anfänger Grenzen setzen; ein knappes Glossar hätte hier gute Dienste getan.

G. Kraepelin (Berlin)

J. R. Norris and M. H. Richmond (Editors), Essays in Applied Microbiology. VIII + 328 S., 98 Abb., 49 Tab. Chichester-New York-Brisbane-Toronto 1981. John Wiley & Sons. £ 19.50.

Diese Sammlung von 11 Übersichten über besonders praxisnahe und zukunftsträchtige Gebiete der Mikrobiologie vermittelt einen ausgezeichneten Einblick in den Erkenntnisstand und die Perspektiven der Produktion von Wirkstoffen und Einzellerprotein. Sowohl biologisch-technologische (Stammhaltung, Selektion, physiologische und genetische Manipulationen, Fermentationstechnik) als auch ökologische (Kontaminationsprobleme, Umweltschutz) und ökonomische Aspekte (Rentabilität) werden berücksichtigt. Haupttendenzen der weiteren Entwicklung wie die Gentechnologie oder die Technik der Zell- und Gewebekultivierung werden von international renommierten Bearbeitern vorgestellt und bewertet. Im einzelnen finden sich folgende Beiträge: A. L. Demain, Angewandte Mikrobiologie — eine persönliche Einschätzung (Theoretische Grundlagen der Überproduktion mikrobieller Metabolite und Enzyme); R. L. Hill, Konservierung von Mikroorganismen (Konservierungstechnik, Kulturensammlung); E. C. Dart *et al.*, Gentechnologie (Methodik der Erzeugung von rekombinanter DNS, Wirts-Vektor-Systeme); J. D. J. Hockenhull, Produktion von Antibiotica durch Fermentation (Penicillin-Produktionsverfahren, u. a.); H. C. Barford, Fermentative Enzymproduktion (Methodik Enzymherstellung einschließlich biochemischer Technologie); J. C. Norris, Produktion von Einzellerprotein (Energiequellen, Applikationsprobleme); C. M. Ignoffo, Mikrobielle Insekticide (Organismen, Produktion und Anwendung); K. J. Seal *et al.*, Biologische Abfallbeseitigung (optimale Bedingungen, Abbau spezieller Materialien); P. D. Walker *et al.*, Bakterielle Produktion von Vakzinen (Bakterielle Toxoide, Antigene); P. D. Stones, Virusvakzine (Herstellung antiviraler Vakzine, Qualitätskontrolle) und J. Melling *et al.*, Schutz vor Kontamination durch Mikroorganismen (Sicherheitsmaßnahmen, biologische und physikalische Kontaminationsverhütung). Der Verzicht auf allzu detaillierte Darstellung der einzelnen Sachgebiete zugunsten einer perspektivischen Übersicht macht den vorliegenden Band auch als Lehrbuch für Studierende der Biotechnologie und Mikrobiologie geeignet.

U. Gräfe (Jena)

A. H. S. ONIONS, D. ALLSOPP and H. O. W. EGGINS, SMITH's Introduction to Industrial Mycology (7th Edition). 398 S., 220 Abb., 22 Tab. London 1982. Edward Arnold. £ 37.50.

Mit der Neuauflage von SMITH's „An Introduction to Industrial Mycology" ist den Autoren zweifellos eine wesentliche Verbesserung und Erweiterung der alten Auflage gelungen.

Nach kurzer Einleitung über Morphologie, Ökologie, Physiologie und Nomenklatur der Pilze werden in einem knappen Überblick die für die Industrie (und sicherlich nicht nur dort) wichtigsten Vertreter der Zygomyceten, Ascomyceten, Hefen und hefeähnlichen Pilze, Basidiomyceten und Fungi imperfecti vorgestellt. Mit Hilfe der Bestimmungsschlüssel, guter Photographien und überzeugender Zeichnungen, die das Wesentliche aus den Photos exakt herausarbeiten und in der alten Auflage noch völlig fehlten, sollte es leicht möglich sein, die aufgeführten Pilze zu bestimmen. Den Gattungen *Aspergillus, Penicillium* (und Verwandte), die unzweifelhaft die größte Bedeutung für den industriellen Mykologen besitzen, wird dann besonders viel Raum gewidmet. Diese Kapitel sind beeindruckend in ihrer Klarheit und ebenfalls ausgezeichnet bebildert.

Die folgenden Kapitel beschäftigen sich mit den wichtigsten Labormethoden sowie mit Schadwirkungen durch Pilze in der Praxis und deren Verhütung. Die industrielle Nutzung von Pilzen wird auf 11 Seiten nur kurz gestreift und somit dem Titel des Buches eigentlich nicht gerecht. Dies entspricht jedoch der Absicht der Autoren, dem Nicht-Spezialisten im Industriebereich zu helfen, die für ihn wichtigen Pilze zu identifizieren. In diesem Sinne möchte man auch das Glossar und die Auflistung von Nährmedien am Ende des Buches nicht missen.

Allen Kapiteln ist reichlich weiterführende Literatur angefügt. Eine weitere Verbesserung wäre gewesen, wenn die Autoren das Kapitel der alten Ausgabe über grundlegende mykologische Literatur übernommen und auf den neuesten Stand gebracht hätten.

Das vorliegende Buch besticht durch seine Klarheit und Handhabbarkeit und kann uneingeschränkt allen, die sich mit Pilzen beschäftigen, empfohlen werden. H. DILL (Berlin)

T. A. ROBERTS, G. HOBBS, J. H. B. CHRISTIAN and N. SKOVGAARD (Editors). Psychrotrophic Microorganisms in Spoilage and Pathogenicity. XIX + 529 S., 92 Abb., 193 Tab. London-New York-Toronto-Sydney-San Francisco 1982. Academic Press. $ 49.50.

Das Buch beinhaltet die Vorträge eines (des ersten) Symposiums über Wachstum und Stoffwechsel von Mikroorganismen bei Kühltemperaturen, das im Sommer 1980 vom International Committee on Food Microbiology and Hygiene (einem Gremium der International Union of Microbiological Societies) in Dänemark veranstaltet wurde.

Insgesamt handelt es sich um 52 Beiträge über mikrobielle Aktivitäten bei niedrigen Temperaturen. Die Rolle der psychrotrophen Mikroorganismen (also derjenigen, die zwar ihr Temperaturoptimum im mesophilen Bereich haben, aber trotzdem bei Temperaturen unter 7—10 °C noch mit beachtenswerten Generationszeiten zu wachsen vermögen) steht im Mittelpunkt der Betrachtungen.

Im ersten Hauptteil sind einige wenige Beiträge mit Aussagen zu allgemeinen Fragen (Definitionen, Enzymcharakterisierung, Plasmidtransfer, *etc.*) vereinigt.

Der zweite Hauptteil enthält die Vorträge über Nahrungsmittelverderb bei niedrigen Temperaturen. Im einzelnen sind sie gruppiert in Milch und Milchprodukte, Fleisch und Fleischprodukte, Fisch und andere Nahrungsmittel (Früchte, Gemüse, Bier). Eine Vortragsgruppe in diesem Hauptteil beschäftigt sich mit dem Einfluß des Mikroklimas (Gase, pH, *etc.*) auf die Aktivitäten psychrotropher Mikroorganismen.

Die Beiträge des dritten Hauptteiles vermitteln schließlich zahlreiche Erfahrungen mit dem psychrotrophen Verhalten einiger pathogener Bakterien. Im Vordergrund stehen dabei *Yersinia enterocolitica* und *Clostridium botulinum*. Ein Beitrag befaßt sich mit Leptospiren.

Ein Sachverzeichnis ermöglicht eine schnelle Orientierung über spezielle Fragen.

U. TAUBENECK (Jena)

A. H. ROSE (Editor), Economic Microbiology Vol. 6: Microbial Biodeterioration. XIV + 516 S., 98 Abb., 46 Tab. London-New York-Toronto-Sydney-San Francisco 1981. Academic Press. £ 38.60, $ 79.50.

Die Materialzerstörung durch Mikroorganismen gewinnt immer mehr an Bedeutung. und ihre ökonomische Dimension wird immer klarer erkannt. Es ist daher verständlich, daß eine umfassende Darstellung des Problems in die Reihe „Economic Microbiology" aufgenommen wurde.

Während die Prozesse des mikrobiellen Abbaus natürlicher Substrate als essentielle Teile natürlicher Stoffkreisläufe (oder auch induzierter Rezyklisierungsvorgänge) hier nicht zur Debatte stehen, werden die eigentlichen mikrobiellen Zerstörungsprozesse, die zur Verschlechterung der Qualität und damit des Gebrauchswertes von Materialien und Gegenständen führen, ausführlich behandelt.

In 13 Kapiteln stellen Experten die wichtigsten mikrobiellen Zerstörungserscheinungen, ihre wissenschaftlichen Grundlagen, ihre ökonomischen Auswirkungen und die Möglichkeiten ihrer Bekämpfung dar. Im einzelnen werden in der Reihenfolge der Kapitel behandelt: Holz, Gestein, Wolle, Felle und Häute, Metalle, Gemälde und Skulpturen, Tabak, Brennstoffe und Öle, Latexfarben, Gummi, Arzneimittel und Kosmetika, Plaste.

Der Herausgeber der Publikationsreihe, A. H. ROSE, hat es auch bei diesem Band verstanden, kompetente Autoren für einen fundierten Gesamtüberblick zu gewinnen. Wenn auch deutlich wird, daß das qualitative und quantitative Ausmaß der mikrobiellen Korrosionserscheinungen bei den einzelnen Materialien bis jetzt noch sehr unterschiedlich untersucht ist, so stellen die hier zusammengetragenen Fakten doch einen wertvollen Fundus dar, der das gegenwärtige Wissen widerspiegelt.

U. TAUBENECK (Jena)

R. F. SCHLEIF and P. C. WENSIK, Practical Methods in Molecular Biology. XIII + 220 S., 49 Abb. Berlin-Heidelberg-New York 1981. Springer-Verlag. $ 32.20.

Der Band ist aus einer Sammlung von Laboratoriumsmethoden hervorgegangen, die im Laufe der vergangenen 15 Jahre von den Autoren ständig verbessert, modernisiert und durch neue bereichert worden ist. Das Resultat ist ein nützliches Buch, das ein umfangreiches Arsenal von molekularbiologischen Arbeitsvorschriften einschließlich leicht verständlichen background-Informationen enthält. Seine 7 Kapitel beziehen sich auf Arbeiten mit *E. coli*, dem Bakteriophagen Lambda, Proteinen, Nucleinsäuren und rekombinanter DNA. Ein Anhang, der Rezepte für diverse Medien enthält, eine Bibliographie, die Arbeiten bis 1980 umfaßt, und ein sorgfältig zusammengestellter Index beschließen das Buch. Alle Seiten haben einen breiten Rand, der Raum für individuelle Notizen bietet. Die Autoren versichern im Vorwort, daß sie nur Prozeduren aufgenommen haben, die sie selbst wiederholt angewendet und die sich als hochzuverlässig erwiesen haben. Dies macht das Buch zweifellos besonders für Anfänger attraktiv. Es zwingt andererseits zur Opferung wichtiger Techniken, die gegenwärtig eine große Rolle spielen. So wird z. B. bezüglich der Anwendung von Transposons und Durchführung von Sequenzanalysen nur auf die Literatur verwiesen. Ebenso verbergen sich unter der Überschrift „Pedigrees and genetic maps" leider nur Hinweise auf die Literatur. Im ganzen scheint mir dies ein gutes Buch für alle zu sein, die gerade beginnen, grundlegende molekularbiologische Techniken anzuwenden. Für viele spezielle Probleme wird man das Buch nur im Verein mit verwandter, aber schärfer fokussierter Literatur über molekularbiologische Laboratoriumsmethoden verwenden können. Es enthält einen ärgerlichen Fehler auf S. 137, wo gleich zweimal steht, daß DNA-Insertion in den *Pst* I-Ort von pBR 322 das Gen für Tetracyclinresistenz inaktiviert. Dies ist gravierend, weil sich unmittelbar daran die Prozedur für die Anreicherung von rekombinanten Plasmiden mit Tetracyclin und Cycloserin anschließt.

H. MALKE (Jena)

A. J. SHATKIN (Editor), Initiation Signals in Viral Gene Expression. 207 S., 30 Abb., Tab. Berlin-Heidelberg-New York 1981. Springer-Verlag. DM 85,—.

Die Untersuchung molekularer Mechanismen der Virusreplikation hat in den letzten Jahren viel zum Verständnis der Genexpression in eukaryotischen Zellen beigetragen. Die Entdeckung des „Cap" durch SHATKIN und Mitarbeiter warf die Frage nach der Funktion dieser Gruppe auf. Einige Antworten werden in den Kapiteln von KOZAK und von KRUG gegeben. KOZAK entwickelt sein „Scanning Modell" der Proteinsynthese-Initiation auf der Grundlage von *ca.* 100 mRNA „leader" Sequenzen. KRUG beweist, daß Influenzaviren für ihre Primärtranskription Primer benötigen, die sie durch spezifische Spaltung zellulärer mRNA herstellen. Mit einer gesetzmäßigen Fehlregulation virusinfizierter Zellen befaßt sich PERRAULT in dem längsten Artikel, mit der Entstehung und Replikation viraler DI-Partikeln. Er präsentiert ein Modell für die Bildung von DI-Partikeln während der RNA-Replikation, die einer „copy choice" Rekombination entspricht. Die weiteren Beiträge behandeln die Kontrollregionen für eukaryotische RNA-Polymerasen II und III (SHENK), die Rolle des „Splicing" für die Regulation der viralen Genexpression (FLINT) und das SV40 T-Antigen, ein Protein das die frühe DNA-Transkription unterdrückt, die virale DNA-Replikation initiiert und auch die zelluläre DNA-Synthese stimuliert (TIJAN). Das Buch bietet einen soliden Überblick über den gegenwärtigen Stand einer Forschung, die noch lange nicht abgeschlossen ist.

C. SCHROEDER (Berlin)

M. P. STARR, J. L. INGRAHAM and A. BALOWS (Editors), Annual Review of Microbiology Vol. 35. VIII + 684 S., 59 Abb., 21 Tab. Palo Alto 1981. Annual Reviews Inc. $ 21.0.

Der autobiographische Beitrag dieses Bandes stammt von H. B. WOODRUFF, der sich, ausgehend von einer bodenmikrobiologischen Grundlage, zu einem der führenden Industrie-Mikrobiologen entwickelt hat und den mit der Antibioticaproduktion gekoppelten enormen Aufschwung der modernen technischen Mikrobiologie aktiv mitgestaltete. Dieser Bericht über eine steile Karriere in der angewandten Mikrobiologie enthält viel Lesenswertes über die Notwendigkeiten und Möglichkeiten des engen Zusammenwirkens von Forschung und Produktion, wenn es zu einer erfolgreichen ökonomischen Realisierung wissenschaftlicher Ergebnisse kommen soll.

Weiterhin enthält der Band 21 Übersichten über unterschiedlich umfangreiche Gebiete aus allen Teilen der Mikrobiologie. Artikel über echte Brennpunkte der Entwicklung stehen neben solchen, die dem derzeit vielleicht weniger beachteten, aber für die disziplinäre Entwicklung der Mikrobiologie ebenso wichtigen Erkenntnisfortschritt dienen. Hier können nur die Themen mit ihren wichtigsten Bezügen genannt werden.

Zur Systematik und zur Darstellung spezieller Leistungen der Mikroorganismen gehören die Aufsätze über die Biologie der hämotrophen Bakterien (Kreier, Ristic), die Taxonomie der gleitenden Bakterien (Reichenbach), die gleitende Beweglichkeit von Prokaryoten (Burchard) und die Biologie der Hyphomicrobien und anderer gestielter und sprossender Bakterien (Moore).

Unter vorwiegend morphologisch-funktionellen Gesichtspunkten wird die Glycocalyx der Bakterien besprochen (Irvin, Cheng).

Aspekte der Pathogenese stehen bei den Arbeiten über die Pathogenese und Immunologie von *Treponema pallidum* (Fitzgerald) und über bakterielle Nierenkrankheiten von Salmoniden (Fryer, Sanders) im Vordergrund.

Der Problemkreis Pflanzenkrankheiten ist durch 4 Beiträge repräsentiert: Interaktion von Bakterien und Pilzen mit Lektinen und lektinartigen Substanzen (Pistole), Entwicklung von Nematoden als Pflanzenparasiten (Maggenti), Schwerpunkte der biologischen Kontrolle von Pflanzenkrankheiten (Schroth, Hancock) und Plasmide, die pflanzliche Hyperplasien bewirken (Nester, Kosuge).

Stark ist diesmal auch die Immunologie vertreten mit den Themen: Psychoneuroendokrinologische Einflüsse auf die Immunantwort (Solomon, Amkraut), Antigene Charakterisierung von Viren durch monoklonale Antikörper (Yewdell, Gerhard), Wirtsabwehrmechanismen an mucösen Oberflächen (McNabb, Tomasi) und Immunbiologie der Karies (McGhee, Michalek).

Genetisch-molekularbiologische Ergebnisse werden im Zusammenhang mit den Gebieten Evolutionäre Bedeutung zusätzlicher DNA-Elemente bei Bakterien (Campbell), Genetik und Regulation der Stickstoff-Bindung (Roberts, Brill), Genetische Studien mit bakteriellen Protoplasten (Hopwood) und Translationsinitiation bei Prokaryoten (Gold, Pribnow, Schneider, Shinedling, Singer, Stormo) behandelt.

Die physiologischen und biochemischen Aspekte der aeroben Wasserstoff-oxydierenden Bakterien werden von Bowien und Schlegel dargestellt.

Schließlich stammt aus der Feder von Alexander eine grundsätzliche ökologische Betrachtung zu der Frage, warum mikrobielle Räuber und Parasiten ihre Opfer bzw. Wirte nicht vollständig eliminieren.

Annual Review of Microbiology 1981 bietet wie alle vorangegangenen Bände gut aufbereitete, unentbehrliche Information über wichtige Aspekte des mikrobiologischen Fortschritts.

U. Taubeneck (Jena)

J. N. Strathern, E. W. Jones and J. R. Broach (Editors), The Molecular Biology of the Yeast *Saccharomyces*, Monograph 11 A: Life Cycle and Inheritance. 751 S., 87 Abb., 43 Tab. Cold Spring Harbor 1981. Cold Spring Harbor Laboratory. $ 90.00.

Die Herausgabe des ersten von zwei Bänden über die Molekularbiologie der Bäckerhefe ist seit dem Erscheinen von „The Yeasts“ (1971) die erste zusammenfassende Darstellung der auf dem Gebiet der Molekularbiologie und Genetik der Hefen inzwischen erzielten experimentellen Befunde sowie deren theoretische Verarbeitung. Von führenden Experten erfolgte eine dem neusten Stand (Literatur bis 1982) entsprechende eingehende Abhandlung zentraler Fragen des Lebenszyklus und der Vererbung von *Saccharomyces cerevisiae*. Die ersten fünf Beiträge geben einen Überblick über die Hefe als Objekt, die Genomstruktur und Kartierung sowie cytologische und genetische Analyse des Lebenszyklus von *S. cerevisiae*. Drei weitere Übersichten befassen sich mit aktuellen Fragen der Sexualität, insbesondere der Rolle der Sexualpheromone, der Genetik der Paarungstypen sowie Meiose und Ascosporenbildung. Die Mechanismen der Genkonversion und mitotischen Rekombination sowie Mutagenese- und Reparaturprozesse werden gebührend abgehandelt. Die sich erst in jüngster Zeit schnell entwickelnde Genetik extrachromosomaler Faktoren wie Killerfaktor und 2 μm-Hefeplasmid-DNA wird in übersichtlicher Form dargestellt. Den Abschluß bilden zwei Beiträge zu Fragen der Mitochondrienstruktur und -genetik, einem hochaktuellen Gebiet der Hefegenetik. Außerordentlich wertvoll sind die in mehreren Appendices erfolgten Darstellungen der genetischen Nomenklatur, der neusten Genkarte sowie der z. Z. bekannten Gene von *S. cerevisiae*, der Genprodukte sowie Zusammenstellung der Gene nach Funktionen. Insgesamt ist dieser Band eine außerordentliche Bereicherung der Hefe-Literatur.

H. Weber (Jena)

FSC
www.fsc.org
MIX
Papier | Fördert gute Waldnutzung
FSC® C083411

Zeitfracht Medien GmbH
Ferdinand-Jühlke-Straße 7
99095 Erfurt, Deutschland
produktsicherheit@kolibri360.de